A Revolution in Egg Production

An Exposition of Practical, Tested and Successful Methods of Continuous Laying

by George G. Newell

with an introduction by Jackson Chambers

Self Reliance Books

Get more historic titles on animal and stock breeding, gardening and old fashioned skills by visiting us at:

http://selfreliancebooks.blogspot.com/

Introduction

I am pleased to present yet another title on Poultry.

The work is in the Public Domain and is re-printed here in accordance with Federal Laws.

As with all reprinted books of this age that are intended to perfectly reproduce the original edition, considerable pains and effort had to be undertaken to correct fading and sometimes outright damage to existing proofs of this title. At times, this task is quite monumental, requiring an almost total "rebuilding" of some pages from digital proofs of multiple copies. Despite this, imperfections still sometimes exist in the final proof and may detract from the visual appearance of the text.

I hope you enjoy reading this book as much as I enjoyed making it available to readers again.

Jackson Chambers

Kellerstrass Farm
Arthur Oscar Schilling
1907

2

Outside View of Egg Factory at Night

Without is Dark and Dreary — Within is Light and Cheery.

Foreword

It is the author's intention, in this work, not only to supply detail knowledge of how to care for poultry, but also to offer the reader something on the order of a post-graduate study.

Details will be gone into because they play a part, and have a bearing, on the new viewpoints presented.

"Coming events cast their shadows before them," and the observant poultryman can often visualize conditions in such a manner as to sense the necessity of changing diet, or surroundings, before the cause of such a necessity has worked to his detriment.

Heredity and environment are two big factors in egg production; but environment has, in the author's opinion, fully as much influence on production as heredity.

A third factor exists, which heretofore has not been recognized as essential, which is under our control, this factor being the length of the hen's business day.

The wild jungle fowl, through changed conditions, and the accumulation of the results of these conditions as transmitted by heredity, has been bred and fed to lay a greatly increased average number of eggs annually; and in the same manner, and for the same reasons, we can, by providing still better environments, further materially increase production— thereby adding to the pleasure and profit of poultry keeping as a business proposition.

THE AUTHOR.

PART I

An Outline of Present Conditions

CHAPTER I—FEEDING AND CARE

CHAPTER II—HOUSING AND APPLIANCES

CHAPTER III—BROODING, BREEDING AND YARDING

CHAPTER IV—PRODUCTION AND CARE OF EGGS

PART II

The Revolution and Its Results

CHAPTER V—BASIC THEORIES

CHAPTER VI—REMARKABLE RESULTS

CHAPTER VII
PRACTICAL APPLICATION OF PROVEN THEORIES

CHAPTER VIII—CONCLUSIONS

PART I.

An Outline of Present Conditions

CHAPTER I

Feeding and Care

Care and Feed in General

One of the greatest requisites for success in caring for poultry, and obtaining satisfactory results, is regularity.

Nothing we can do upsets a laying hen's domestic economy so much as to be left without an expected meal, or to be left without water.

Attendance must be constant. Some flocks are kept in an alternate condition of feast or famine, and the owner has no right, or reason, to expect good returns from such treatment.

We can all see and observe how wild birds are scared away and made to desert their nests, if they are in any way disturbed, even if this disturbance happens when the birds are entirely out of sight.

Domestic fowls in the same manner, only possibly modified in degree, have a knowledge and sense of whether conditions surrounding them are such as to be conducive to safety in laying.

The effect of this trait is easily observed, and any doubter can satisfy himself of its truth, when sudden changes are made in quarters of laying hens, even if the changes are made for the better.

The hens or pullets, for the time being, are disturbed by the change in appearance of their surroundings, and laying stops in a greater or less degree, according to the mental attitude of the birds toward these changes.

In the same manner, laying is affected by the birds' sense of security and quiet. If they are continually frightened, or disturbed, they will not lay as well as if they feel contented and satisfied. In some subtle or mysterious manner, it will be found that poultry can mentally size up, and figure out, their surroundings to the profit or loss of their owners.

Some poultrymen, by their manner or methods, are never able to give their birds a satisfied and "comfy" feeling, while others, with apparently less effort, are quite at home among their birds, and the birds are quite at home with them.

Balancing the Ration

To the poultryman keeping only a few birds, it is sometimes quite a problem to secure a variety of separate grains to fit the needs of his flocks. "Variety is the spice of life" and laying hens take no exception to the proof of this saying.

The poultry business has made such strides in the past few years, that there are several brands of poultry feeds put up in whole and ground grains in all parts of the country. These mixtures are mainly scientifically mixed in proper proportions for a theoretically balanced ration; and they cost no more, if as much, as the same ingredients can be secured for, separately, by anyone desiring to mix them himself.

It is an open question, in the present condition of feed supplies, whether even the larger poultrymen cannot save time and money by purchasing their feeds already mixed, for the bulk of their feeding.

Whether mixed feeds or separate grains or meals are purchased, the fowls themselves will balance their own rations, to a great extent, if given an opportunity to do so.

They will do this by choosing the grains they require out of the mixed grains, and leaving the others untouched, even to the extent of going hungry sometimes, rather than eat such grains as they do not require or fancy.

If grains are fed separately, we therefore may often be forcing the fowls to either eat what they do not desire or require, or go hungry; and we may, so to speak, be forcing our judgment of what is a balanced ration upon the birds.

No one variety of grain contains the necessary food elements in the right proportions for our purpose. If poultry are forced to sustain themselves on a restricted diet, they will have to pass through their systems much useless material (measured by their needs) in an attempt to obtain sufficient of the necessary elements which they require; for instance, hens fed principally on corn will put on fat to such an extent as to be unable to produce eggs. The same, or less, expenditure for a more balanced ration would produce profitable results.

Our judgment in such matters is not at all comparable to that of the birds, "Ask the birds, their judgment is good."

Another error we are likely to fall into is that a balanced ration for today and today's conditions, will be a balanced ration for next week and next week's conditions; when, as a matter of fact, the requirements may be entirely different.

Still another argument in favor of mixed grain of the desirable kinds, is that the requirements of the individual birds in the same flock may vary every day. A hen who has been broody for some time, and has been neglecting her feeding in consequence, will require more fattening foods than a hen in laying condition; and if these fattening foods are not available, such a hen will remain out of laying condition indefinitely.

For the reasons touched upon in the foregoing, let me here suggest an efficiency method for determining a balanced ration for those making a business of keeping poultry, even if mixed grains are purchased. The method is this: Choose a time when the fowls are fairly hungry, say in early morning, and place in separate troughs a definite quantity of separate grains. Suppose we use two pounds of Kaffir corn, two pounds of wheat, two pounds of cracked corn, two pounds of oats, and two pounds of sunflower seed, etc. After these grains have been before the hens some time, we will find that some varieties have entirely disappeared, while some have scarcely been touched, and others have been cleaned up in varying extent.

The hens will thus tell us what they would like added to their rations, and we can either feed such grains as separate extra feeds, mix more of such grains in with our mixed feed, or, if we use hoppers, we can increase the quantity of such grains in the hoppers.

We can make this test as often as we think necessary,

say once a week, once semi-monthly, or once a month; and we will thus be in a position to meet the requirements of a ration balanced by the hens themselves. "Ask the birds, their judgment is good."

Conditions Should Be Watched and Noted

To be a successful poultryman, there are no qualities more necessary or valuable than the power of observation, and decision for quick action in correcting conditions when anything wrong is observed.

After disease takes a thorough hold on a fowl, to such an extent as to sap its vitality, the ax is about the best remedy one can use, generally speaking; but, if conditions were noted earlier, the fowl could have been saved and brought into good condition quickly.

Sick fowls should be separated from the well ones at once—if for no other reason than that they should not use the same drinking fountains, and thus contaminate the water.

The droppings are a valuable indication of trouble. If they are without form, or greenish, they suggest at once something wrong in the feeding, or the necessity of charcoal or grit. If the droppings are of a consistency to hold their shape without being too solid, it is a good indication.

For excessive looseness of the bowels the old fashioned Douglas Mixture is a quick remedy, and seems to act as a tonic. This mixture is greatly in favor among English poultrymen. The formula for Douglas Mixture is eight ounces of Sulphate of Iron (commonly called copperas or green vitrol), and one ounce of Sulphuric Acid, dissolved in two gallons of water.

This mixture should be kept in glass or stoneware, and should never be placed in metal receptacles either in the stock solution, or when placed before the flock. A teaspoonful to about a quart of drinking water will act as a tonic, and may be given to fowls which are well and to the sick birds.

The regular drinking supply may be taken away temporarily, and the earthenware vessels substituted containing the diluted Douglas Mixture.

Looseness of the bowels is not always "cholera," but may be caused by faulty diet. The writer some years ago had a flock of young Plymouth Rocks, three months old, which, unaccountably and all at once, developed great looseness of the bowels. Two birds were found dead, and some others in bad shape. The flock had apparently enjoyed good health, and even as bad as they then looked, did not have a diseased appearance. Thinking back in an attempt to discover the cause of the trouble, he remembered they had drank great quantities of water the day before; and he then remembered that the family had partaken of salted herrings, and, on inquiring, found that the bones and other parts of the herrings, which were left over, had been thrown out to the fowls; hence, exceeding thirst, inordinate drinking of water, and the fatal looseness of the bowels. A hurried trip to the drug store for material, and Douglas Mixture made and given to the fowls, saved every one of the balance of the flock. Two days afterward not a sign of the trouble was to be seen. Without quick action every bird would have been lost.

Simple colds may be caused by openings creating a draft, and these openings should be attended to at once. Neglected colds and contamination of the water soon lead to roup, and demoralization of the flock.

If eggs are laid with soft shells, or a tendency develops for egg eating, look to the supply of oyster shells in the hoppers.

The quality and quantity of green feed has a big influence on digestion, and its derangement. Cabbage is a good winter feed, but if overfed to fowls is likely to scour them.

What to Feed

For the small poultryman it is sometimes quite a problem to decide what to feed. The grains or meals available in the local market may be very restricted, and if he confines himself to such a market, it may seriously affect his returns, because of the poor or unbalanced rations fed.

Commercial poultry food can, however, be secured now in all parts of the country; and, in the writer's opinion, should be used—especially by the small producer, who either cannot secure the necessary ingredients to do his own mixing, or has not the knowledge to decide what is needed in the ration.

Broadly speaking what are necessary to feed are the following:

A variety of grains for the fowls to choose from.

A good dry mash, consisting of clover or alfalfa meal, bran or middlings, beef or fish scrap (or both) ground oats, cornmeal, linseed meal, etc.

Good sharp grit.
Oyster shells to supply lime.
Charcoal,
Beef scraps.
Green feed of some kind.
Water.

Buckwheat and sunflower seeds will be valuable to have on hand, as extra feeds, and will be relished by the fowls, especially in winter. They seem to enjoy these feeds in greater quantities than they are generally found, in the commercial mixed grains.

Feeding must be governed more or less by the breed handled. The active breeds, such as Leghorns, can use a greater proportion of fattening foods than the heavier breeds, without getting out of condition.

Green feed and its furnishing is one of the greatest problems, especially to the city dweller or suburbanite.

One has no right to trouble one's neighbors by allowing their poultry to run loose over their lawns or gardens. Mischief done in this manner by the fowls can not be made good; and such results are not conducive to neighborly good feeling.

If attention is paid to this matter, however, fowls can be confined in small yards the whole year, without bothering the neighbors and without suffering from lack of green foods.

Cabbage, yellow turnips, rutabagas, or beets, etc., can be purchased anywhere, and these are relished by the fowls. In a large plant, it will pay to grind up such vegetables as beets and turnips and feed them in the mash.

Sprouted oats is king of green foods for poultry, and nothing is so much relished by them. Good oat sprouters can be made or purchased. Oats can be sprouted in every season of the year. In the winter, however, heat must be supplied when the temperature gets too low, as the oats will be spoiled and the sprouts will not grow if the oats get chilled.

In very warm weather the oats are likely to mold and thus become unfit for feeding, besides stopping the sprouting.

Chemicals are advocated to stop this molding; but personally the author has not felt like using the chemicals advocated to feed to his flock. He has found, however, that a little air slacked lime put in the water, when sprinkling the oats, puts a stop to the mold, and is not a bad thing for the flock.

The grit used must be sharp and hard, as this material is what furnishes the birds with a substitute for teeth, to grind their food. The food, when it reaches the gizzard, is separated and prevented from getting into a mass, and is ground up between the hard particles of grit.

It is hard to realize what powerful action the gizzard is capable of. In the year 1899 the writer was talking to a butcher about a lot of fowls which had been confined for some days, and seemed to be off their feed. The butcher was told they needed grit to sharpen their appetites, but there was no grit at hand. The writer suggested that some glass be pounded up for them, to which suggestion the objection was offered that it would certainly kill the fowls. The glass was fed to them, however, and their appetites at once improved. The butcher had expected to surely lose some birds.

When the hens were dressed, he said, "Now I am going to see what happened to that glass," and when the gizzards were opened he was surprised to find nothing but round and smooth crystals, in place of the sharp pointed particles of glass which had been fed to the hens.

Under Feeding

It takes feed and plenty of it to get eggs. The reason most people get poor, and unsatisfactory, egg yields is that they either do not realize how much feed is necessary, or wilfully try to economize by not giving the fowls more feed. This is false economy, as keeping unproductive fowls is an expensive proposition.

Of course, the keeping of a few fowls fed on table scraps is not felt as an expense. On the other hand, if they do not produce, the available profit from correct feeding is lost—even if the owner does not realize his loss.

Some, to be sure, may keep a flock of hens for the pleasure of having them around, and for their beauty. The beauty is not spoiled, and the pleasure is not detracted from, but rather enhanced in degree, if they furnish a plentiful supply of eggs for the table—to say nothing of the difference in quality between such eggs, and the eggs which can be purchased at the stores.

Where large flocks are kept, the feeding expense is a heavy item, and if the fowls do not produce satisfactorily they will soon make a heavy drain on the bank account, instead of adding to it.

It has been said that "Man never is, but always to be blest," and many flocks of hens, if they could voice their feelings, would echo the same sentiment by saying, "That they cannot lay, or are always on the point of laying; but fall short because the necessary feed is withheld from them—they are always on the point of laying but don't lay." A balanced ration, **and plenty of it,** would soon make such flocks produce; "Ask the birds, their judgment is good."

Over Feeding.

If a flock of fowls has been reared under generous feeding conditions, there is not much danger from over feeding if the rations are balanced fairly well. Of course, a flock fed on nothing but corn, or other fattening food, would soon get over fat—but still not produce eggs; because some materials necessary to make eggs are always conspicuous by their absence, and the fowls in an endeavor to obtain sufficient of these necessary elements, which are present in some proportion in all grains, over eat of other elements and lay on fat.

With the heavy breeds, it is probably necessary to be more particular as to the amount of exercise forced on the fowls, than with the lighter breeds, such as the Leghorns.

Observation, however, will show that a hen has to be in fairly good condition for continuous laying. The author has used White Wyandottes for the table, which before being dressed were laying, and from which as much as one pound of clear fat was taken before cooking.

An over fat hen will be far more likely to lay than an under fed hen. With the lighter breeds, Leghorns for instance, if the birds are fed regularly, without spasmodic periods of scarcity, there is little or no danger of over feeding on a good balanced ration.

We have kept Leghorns in flocks for five years, which never knew a time when there was not feed and water available, at any hour or minute of their business day, if they wanted it—and they laid **some,** and are still at it. For quantity of feed, "Ask the birds, their judgment is good."

The Hens Enjoying a Night Mash

"Nail Keg" Automatic Feeder
Home Made, and Works Satisfactorily

CHAPTER II

Housing and Appliances

Automatic Feeders

For regular and scientific feeding commend me to the automatic feeders. We read so much in the press and poultry journals of directions of how and when to feed, and the quantity to feed for so many birds in a flock, that the author is not loath to take exception to these notions as being entirely wrong, and will give his reasons for so doing.

These reasons will apply equally as well to the small poultryman, keeping a few hens as a side issue, as to the large poultryman who depends on the business for his livelihood. In the first place, we cannot put ourselves in the place of the hens, and judge how much feed they need. If we attempt to do so, we, in effect, say to a certain individual hen that she needs, say two ounces of grain, at a certain feed; when, if she could speak, she would probably tell us, that on that particular day she needed another half an ounce, without which she could not lay an egg next day. For quantity of feed, "Ask the birds, their judgment is good."

As stated before, regularity in feeding is essentially necessary for success, and with this measured hand feeding a poultryman or his family is tied down to the fowls, or fowls suffer as a consequence at times, when they miss a meal; the small poultryman is away from home, the good wife out shopping, train or street car is delayed, dusk arrives and the fowls go supperless to roost. This results in a sudden and protracted break in egg yield.

The same way with the larger poultryman—he cannot be away on business, and fail to reach home without serious consequences.

With automatic feeders, conditions are entirely different. The fowls have plenty to eat as long as daylight lasts, if only their caretakers have filled their feeders before leaving home. Feeders furnish exercise as well as feed, and the amount of exercise can be regulated by adjustments on most of the feeders offered for sale. The exercise can also be regulated by the quantity of litter, and occasional, or early morning feeds scattered in the litter.

Fowls have got to feel that feed is plentiful and always available, if we wish to obtain a full measure of results. Fowls think, and it will pay to keep them thinking contentedly all the time. If you have any doubts about these matters, "Ask the birds, their judgment is good."

Fussing vs. Economy of Time.

As a broad proposition it may be stated, without fear of contradiction, that the fowls will do as well and will not care a whit, if conditions are right, whether these conditions are brought about with little or much work on the part of the attendant. It will not make any difference, to the egg yield, whether the attendant spends all day furnishing these conditions to a flock of hens, or whether he furnishes the conditions in an hour's time; but it will make a vast difference to the attendant in his economy of labor.

"Time is money," and if time is not wasted in fussing it can be utilized in other affairs, or in increasing the capacity of the plant—if all one's time is put in on the plant.

Every device or appliance, therefore, which tends to economize time, shorten or lessen steps, or helps to prevent any leaks should be made use of. Nowadays Efficiency is the Watchword in any line of business, and there is no reason whatsoever for making the poultry business an exception.

Right tools with which to work cannot be considered as an avoidable expense, but should be considered as indispensible assets in conducting successful operations.

Water Problems

The death knell of many an otherwise promising poultry business, is seen in the way water is **dished** out to the fowls.

Sometimes water is carried to the flock in shallow dishes or trays, and the fowls walk all over the dishes, **foul** the water, and make it unfit to drink in a few minutes. Probably, also, the dish is upset in short order; and then there is no water where the water ought to be, and the fowls are soon thirsty.

'A Revolution in Egg
Production"

Points the practical
w a y s to increased
egg yields i n a l l
seasons.
Yours very truly,
Geo. G. Newell

Water—the Cheapest and Also the Most Neglected Necessity

Water in pails or deep vessels is some improvement, but this also soon gets foul, and the fowls become thirsty, for the reason that, when used for a little time, the water gets down so low in the pail, or other vessel, that they cannot reach it.

These open vessels are also quickly frozen over in cold weather, thus cutting off the supply.

Automatic drinking fountains, especially those constructed in recent years, so as to prevent damage to the fountain from freezing, are a great convenience.

The handiest of these are filled like an ordinary pail, and then turned over on their sides.

A Handy Automatic Water Fountain

The poultryman, with such fountains in use, can fill them up once a day and be certain that his fowls are not thirsty at any time. In well constructed houses, water will not freeze in such fountains at an outside temperature of zero or a little below. In extreme temperatures a pail of hot water can be taken, to the poultry house, during the day, the fountain set upright and the hot water poured in, and the fountain again turned on its side.

One of the biggest "chores" in the poultry business is the furnishing of water, and every convenience should be made use of to lessen this work. The water also will be found to be a prolific cause of the spread of some of the most dangerous diseases of poultry.

For this reason, all fowls suffering with colds, roup, etc., should be immediately separated to prevent contagion or infection from spreading to the balance of the flock.

Permanganate of Potash is highly recommended for putting in the water, to assist in the cure and prevent the spread of colds, roup, etc. Sufficient Permanganate of Potash may be dissolved to turn the water to a decided wine color. In case of roup or kindred diseases, the head and eyes can be swabbed with a solution of Permanganate of Potash with good results. About as much as will stay on a dime will be enough for three gallons of drinking water; and the same quantity in a quart of water will be about right for a washing bath.

The poultryman handling large flocks would be short sighted if he did not, by some means, either by pumping under pressure, or by using a natural fall, provide a continuous supply of fresh water, thus promoting the health of his flocks, reducing the labor item in one of its most arduous tasks, and insuring against infection through the water supply, if disease should gain a foothold in his flocks.

Housing, Ventilation and Light.

The much discussed question of open-front vs. closed houses is still an open one. Like most issues of this kind, common sense and the real merit of the question lie in middle ground.

All are agreed that the housing should be tight on all sides except one, and that the ventilation, anyway in winter, should be from one side only to prevent drafts.

Provision should be made, however, to insure plenty of circulation of air in the extreme temperatures of summer, otherwise the fowls will suffer greatly.

The writer has in mind at this writing, a poultry house in Michigan one hundred feet long and fourteen feet wide, which he saw in the summer of 1912. He visited the owner of this house, and the birds looked beautiful and white, and seemed to be in good condition; but they suffered and panted with the heat, and all had their wings spread out. The birds had been kept indoors because he thought it was cooler indoors than out, because of the reflection of the sun on the sandy soil. The air was stifling in the house, as it had a southern exposure and the glaring sun on the front made the air on the inside an extreme temperature, from which there was no escape. He suggested to the owner that a hole be cut through the rear wall, and that the boards be cut in such a manner that they could be battened together to form a door, to close the opening at will.

The effect when the first piece of board was taken out was instantaneous, and the relief was sensed by the birds at once; they all got in line of the breeze to enjoy it. "Ask the birds, their judgment is good."

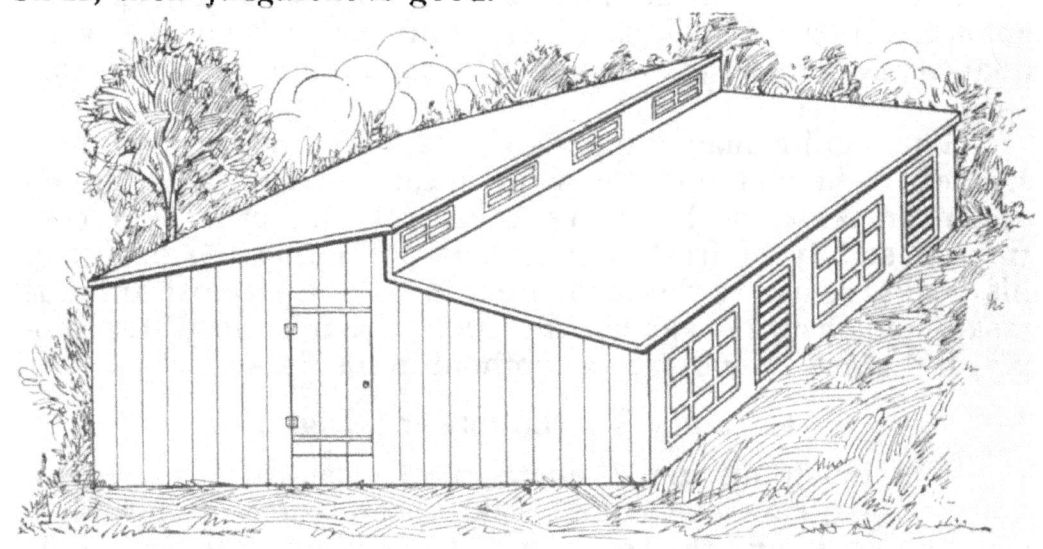

Double Pitch Monitor Roof Poultry House
Showing Slatted Ventilators and Windows in Monitor Roof

The extreme temperature had already resulted in the death of a half dozen birds. We found one hen which seemed about to expire, took her outside, laid her on the grass, and, after pouring cold water over a shaded spot laid her there to cool off. She revived and next day had fully recovered.

The same openings in winter time would be fatal to the health of the flock. Good judgment must be used in these matters. An opening in the roof even, may be a good thing in the summer time, if it is made in such a manner as to be protected by a slatted ventilator, to prevent rain and wind from beating in. Such a ventilator takes the heated air right out of the house.

The house needs plenty of light and sunshine in addition to ventilation. It needs the light at all seasons, and the ventilation in varying degree according to the seasons, and the temperatures of the seasons.

Muslin covered openings are recommended for the double purpose of light and ventilation. However, they are not an unmixed good and are not as effective for these purposes as they are recommended to be. They soon get dusty, so that they do not ventilate freely, and in the course of a season get so dirty that they shut off very much light. The latter objection is not so great in the long days of summer as in the short days of winter.

Openings or windows protected by hinged slatted doors, to keep out the weather and prevent direct air currents, are much more effective for ventilation, and can be fastened back and the windows closed, when not needed as ventilators.

In the summer, if the house gets overheated because of the glare of the sun on a large amount of glass, some of the glass can be washed over with a solution of Whiting and water, and this will protect from the sun's glare without excluding much light.

As to the open front, this is a good summer proposition, if the top of the front opening is below the level of the roosts, and some distance away. If the opening is above the level of the roosts, especially in narrow houses, so that the wind can blow in directly, look out for trouble in fall and spring.

In winter, an open front may prove a delusion and a snare, in extreme temperatures. This danger can be lessened by covering these openings with muslin; but the covering of these openings· will also shut out a great amount of needed light and sunshine.

A closed house is also a source of danger, as fowls need plenty of fresh air. There should always be an opening, such as a window part way open, even in the severest weather. When the outside temperature gets down to zero or below, more ventilation will be secured, and the air will circulate more freely, through a window open three or four inches, than with all the windows on one side open to their full capacity in the summer time.

If you have an open front, and find the birds, on cold days, huddled up to keep warm at the side farthest from your open front, close the openings to a point dictated by observation and judgment, and note the effect on your flock. "Ask the birds, their judgment is good."

A house which is deep from the roost to the front will allow for a greater amount of ventilation, with less danger, than a narrow house.

A very deep house, with a roof of one pitch, will make it necessary to have the front of the house high to admit the

Hen House—Remodeled 1915

sunshine into the back of the house, for sanitation and comfort. When we get a house built too high it will prove cold in winter, and will also be a waste of building material which is used up in enclosing a great amount of dead space which is never used.

We get the same effect of sunshine, without the defects just mentioned, by a break in the pitch of the roof, and making it double pitch with a monitor top.

If we leave the greater span at the back, or roosting side, of the house, we can place windows in the drop or monitor side of the roof. Through these windows the sunshine will be admitted to the back part of a very wide house. These windows can be covered with muslin for summer use, or can be left open entirely in very warm weather. Windows placed in the low front will admit air and sunshine into the front side of the house.

With a little common sense, we will thus have under our control opportunities for plenty of ventilation without drafts, winter or summer.

Dust boxes should be provided in sunny spots, and dusting material should always be kept in good supply in these boxes.

Sifted coal ashes make good dusting material. Wood ashes are not suitable, because of the lye content. Road dust, or good dry earth is good dusting material. Sand may not prove entirely satisfactory for this purpose, because of its being too gritty. Some poultrymen depend on the condition of the floor to furnish dusting material. It is safer, however, to provide separate boxes for this purpose.

The floor and litter should always be dry. At times it is easier to say this than to secure the condition. In a small house, especially, the dampness gets in during the winter, by means of snow on the feet, and by the moisture of the atmosphere condensing in varying degrees on the walls, ceiling or roof. This condensed moisture, when it melts, drops on the floor and dampens the litter. This condensing can be governed to a great extent by ventilation, but cannot be eliminated altogether in extreme temperatures, without providing more ventilation than is good for the fowls.

The roost poles should be sprayed or washed with kerosene, or a mixture of kerosene and disinfectant, every few weeks to destroy mites and other vermin.

Theoretically. the house should also be whitewashed every year or two. Whitewashing has two points in its favor; first the sanitary or purifying effect of the whitewash on the walls, and second, the effect on light conditions, as the white walls will reflect a great amount of light.

Whitewashing a poultry house with a brush will not prove a very pleasant task, however, and if the roost poles are kerosened, and a dust bath is always kept available. a less frequent application of whitewash will prove practical from a sanitary standpoint. A good sprayer makes easy work of whitewashing, and will be as valuable to use in a poultry house for the benefit of the light effect, as for sanitation.

The roost platform should be cleaned off frequently— daily would be good practice. If thoroughly done once a

View Showing the Two Houses Used in These Experiments

week, however, and ashes or dust are strewn over the platforms, with an occasional dusting with air slacked lime, the houses will be kept in a fair and practically sanitary condition.

The litter should also be cleaned out and renewed at intervals.

Nesting material should be cleaned out of the nests at intervals, and a sprinkling of lice powder in the nests, when putting in new material, will help keep the fowls comfortable.

Trap Nesting

Trap nests have their uses, and are a great convenience for certain purposes. There are several very clever trap nest devices on the market. If eggs are desired from certain individual members of the flock only, the only sure method of getting such eggs is by using trap nests, set to lock the individual layer up until released. The eggs can then be marked to identify them.

For the fancier, or anyone desiring special breedings from individual hen's eggs, trap nests are indispensible.

For commercial egg production, some features of trap nesting are a positive detriment. If trap nests are so made that, after laying, the hens can enter another pen where feed, water, etc., are in constant supply, the same as in the pen where they were before laying, there will be no objection to the use of trap nests when egg production is the object. The birds under such conditions can be so marked as to be easily identified; or may be transferred and identified after going to roost. The eggs can be counted to see if the number corresponds to the number of hens, as a check on the correctness of results. Some hens may go on the nests and pass through without laying; therefore this method will not always be positive for exact results. This method will also add some to the expense, in time, if time is valued on a monetary basis.

If regular trap nesting is used, when good egg production is the object, that is, if the hens are locked up until released by the attendant, one of two things **must** happen: First, either the attendant must practically live with the flock, so as to be in a position to release the hens soon after laying; or, second, the capacity of each hen for producing eggs will be cut down in proportion to the length of time that the hens are shut up after laying—thus cutting down their productive powers to that extent, or in other words shortening their business day.

Fowls under such conditions cannot do their best, especially in winter, where every minute of light counts for much. Of course, by the use of trap nests the owner will have the satisfaction of knowing how many eggs each hen lays, and which hens prove the best layers (for the number of hours they have had available for manufacturing those eggs). He will also know, if he stops to think, that his "egg machines" can do better, and each and every one of them probably will lay more eggs without the use of trap nests than with them, unless the attendant is constantly on the job to release the hens. "Ask the birds, their judgment is good."

Incubators or Hens for Hatching?

Whether it will be better to use incubators, or hens, for hatching will depend on circumstances. For the poultryman with only one or two dozen hens, who does not intend to

Pullet House

increase his flock because of space limitations, or other reasons, the hen will probably always be the only practical means of incubation.

For those desiring to increase their flock, or those who already have fair sized flocks, this question will require careful study.

Incubation, in many cases, will be more succesful with hens than with incubators, depending on the make of incubator, the çapacity of the operator for handling the incubator, and the way in which the hens are cared for, etc. This is, of course, presupposing the same hatchability of the eggs used in each case.

An operator, however, can care for an incubator which will hatch as many chicks as thirty hens with about as little effort as caring for one hen; and after hatching there can be no comparison between the work and trouble of caring for the chicks as one lot, and the care and trouble of looking after thirty broods with hens.

The work with hens can be lessened some by doubling up the broods after hatching, or the brooding may all be done by artificial means. Practice will prove that incubator hatched chicks will do better under conditions of artificial brooding than hen hatched chicks. One of the chief reasons for this will be found in the fact that a great percentage, of hen hatched chicks, will be troubled with a legacy of lice, from which pests incubator hatches are free.

The troubles of caring for broods of chicks under hens will be many, and the more hens the more troubles—in proportion to the number of hens.

Under artificial brooding, all the chicks can be kept safe from weather, vermin, etc., and they will rapidly learn to take care of themselves. Under hens they depend on the hen, and at a critical time, when the owner is counting all dangers as practically passed, the hens may leave the chicks to shift for themselves. Hens when left to their own devices, are about the poorest brooding contrivances imaginable. They travel far with the chicks, tire them out, trample on them, smother them, pick at them, and manage to take them through wet grass and leave some of them there to die. Of course, there are exceptions, as some hens will prove very good mothers to the chicks, but the exceptions will prove the rule.

The author had an experience with hens with about fifty chicks, some years ago. The chicks were about four weeks old and were left to shift for themselves during a heavy thunderstorm. The chicks were picked up drowned, half drowned, and chilled, and were scattered all over the place.

Each make of incubator will have instructions from the manufacturer as to how it should be run. Some machines are heated by the circulation of hot water, and others by the circulation of hot air. The heating may be done by means of oil, gas, coal, or electricity.

In trimming lamps, the wick should be rubbed off or be trimmed carefully, and in such a manner that no sharp points are left, especially at the corners of flat wicks. The corners should be slightly rounded. With round wicks they should be rubbed off or trimmed so that no sharp points of flame will appear. It is these sharp points of flame that start the lamps to smoking, and cause danger from overheated lamp founts.

If possible, incubators should be run, for best results, in a basement or some such place where the temperature remains fairly uniform. They can, however, be run successfully in a dwelling, but need far more attention under such circumstances.

The author ran successfully a home made hot water machine, of his own construction, in the year 1892 in a one story building with a flat tin roof. The temperature of the room often went to over one hundred degrees in the middle of the day, and down to nearly freezing at night. At that time incubators were not perfected as they are now, and such as they were, they were expensive luxuries. Several hatches were made in this home made machine for about seven years, but it took constant watching to maintain right temperatures, as it had no regulator, and all the regulating had to be done by guessing at the necessary amount of lamp flame to keep about ten gallons of water at a nearly uniform temperature. The lamp flame was turned up slightly to overcome night temperatures, and turned down in the day time—sometimes turning out the lamp altogether for two or three hours on warm days.

The ideal temperature for the first week is about $102\frac{1}{2}°$ or 103°, for the second week 103° to $103\frac{1}{2}°$. After ten or eleven days the life in the eggs begins to generate bodily heat, and the incubator temperatures may suddenly get too high, unless allowances are made for the generating of this bodily heat in the adjustment of lamps and regulators, or both. Toward the end of the hatch a temperature of 105° may be allowed without harmful results. An increase in the amount of ventilation helps the success of the last few days of incubation.

Success with incubators seems to have become more general since provisions have been made, in most machines, for maintaining the right degree of moisture in the egg chamber.

The first class modern incubator will keep the right temperature like clockwork, if cared for by a person of average intelligence.

Just how much variation in temperature, and other changes of a physical nature—such as moisture, etc.—a hatching egg will go through, and still produce a healthy chicken, is hard to determine.

We know that sustained high temperatures are fatal, that sustained low temperatures mean the death of the germ, that lack of moisture drys down the embryo, so that it cannot develop properly, and causes it to stick to the shell, thus preventing it from freeing itself at the last, and also that an excess of moisture develops the chicken out of proportion, causing it to fill the shell so full that the chicken cannot turn around freely and sufficiently to break its prison walls, when the time of development is up.

A sitting hen gets off the nest to feed, and sometimes stays off the nest, even in cool temperature for two or three hours, and still brings off a goodly hatch of youngsters.

A hen also turns her eggs. Whether she knows why she turns them, or not, is a question. Turning the eggs helps to prevent the embryo sticking to the shell, because the construction of the egg always keeps the live germ uppermost, even before incubation starts, and during incubation the growing embryo remains in a suspended condition in the egg, with the development always on the upper side, and if an egg is turned, the contents will shift so as to keep this same uppermost position, and thus change the position relative to the shell. It is the weakening of these provisions, through staleness of the eggs which causes the "spots" which are thrown out in candling when they reach the market.

The probabilities are that a hen moves the eggs (turns them) in order to get the cooler outside eggs under her, and thus relieve her feverish condition; and in endeavoring to increase her own comfort kills two birds with one stone; first, insures that the eggs are turned, and second, that all the eggs, by being moved and shifted to the center, get an equal share of heat units in the total period of incubation.

It requires a certain number of heat units to develop a crop of corn, or any other grain, and these units must be supplied before a certain limit of low temperature in the season (that is, before frost) prevents further growth.

A hatching embryo, in the same way, must have a certain number of heat units, and must have them without sustained variation above or below certain limits, and also within a certain time. If the heat is maintained at too low an average temperature the chick will not develop sufficiently before it is due to hatch; and if a relative excessive temperature is maintained, the chick is developed before the time of exclusion is due. Normally the chicken will begin to "pip" the shell on the eighteenth day, and will, after "pipping" through in one spot, work all around until the shell is broken in two, on the twenty-first day. When a hatch is late, cloths wrung out of warm water, and placed in the egg chamber, will help the hatch. The author in 1892 had eggs in an incubator which had been shipped from Boston, Mass., to Lead, So. Dakota, and not an egg was "pipped" on the twenty-first day, and the use of damp warm cloths brought off a fair hatch.

To show the variation in moisture which an egg can survive, the author remembers an experience in Hand County, South Dakota, on his mother's farm in 1882. On account of the low temperature of sometimes between 20° and 30° below zero, or lower, a pit was dug out for the house, and an "A" roof put over this pit, with the doors and windows in the south side of the "A". During a heavy thunderstorm the pit was completely filled with water, entirely covering two nests of eggs, and driving the hens off the same. It was over four hours before the water lowered, and the hens went back on the nests, and they brought off good hatches. This was a surprise to us all, as we had given up all idea of seeing any chickens appear.

If eggs are taken just before "pipping" the shell, and placed in a large pan of warm water, the action of the chicks in pecking at the shell causes the eggs to move over the surface of the water, and is a comical sight. Eggs never seem to hatch any poorer for these experiences.

----

*As an illustration of the ability of birds to take care of themselves, and of their knowledge of conditions best suited to their needs, about twenty turkeys roosted on the top of this "A" chicken house every night. Several attempts were made to drive them into more sheltered positions, but without success.

They stayed on top of this "A" house in all weathers and temperatures above zero or thereabouts. If it got below zero, and there was a high wind, they would move; but when not windy they would roost there when it got down to 15° below zero. A drop in the reading of the thermometer to 20° or lower below zero, however, would find the turkeys going to roost on a straw stack about two hundred feet away.

One would think that, knowing the straw stack was there, they would roost there all winter, but they would never seek the shelter of this straw stack, except in stormy weather or extreme temperatures.

The only plausible explanation of this is that, as stated before in this little book, fowls think, and they must have felt more secure on this "A" roof, and determined to stay there, until forced by their sense of danger from cold to seek a better shelter.

CHAPTER III

Brooding, Breeding and Yarding

Brooding Problems

Under hens, brooding, except for supplying proper feed, the requirements of which will not vary much from artificial brooding, resolves itself into care of the hens.

Under artificial brooding other problems have to be faced, the principal of which is supplying artificial heat in proper manner to the chicks.

Artificial brooding may be either indoor or outdoor. Outdoor brooding must almost of necessity be carried on with small flocks, because the capacity of nearly all outdoor brooders is limited to the care of about fifty chicks, at the outside, up to the age when they need no further hovering.

This is a serious objection to outdoor brooding for large flocks, because the work of caring for many brooders in this manner will prove arduous. A more serious objection, however, and one which materially affects the growth and well being of the chicks, will be that, owing to weather conditions, coupled with the fact that probably several hours of day light will pass away before the arrival of the attendant, the chicks will be confined within the limited space of the brooders for too many hours for their welfare.

There are at the present time many good indoor brooders manufactured. Some of these require special provision in the shape of a pit or depression to care for the lamp. Some of the later makes of indoor brooders are made in such a self-contained and safe manner, that they can be placed on any ordinary floor without any special provisions being necessary in the building. These can also be used as outdoor brooders, if thought desirable, or, on special occasions, by placing them in large boxes for protection from the weather.

For a larger poultry business, there are brooders manufactured, and in use, which care for from fourteen hundred to fifteen hundred chickens in one lot. These brooders originated in California, and proved quite a success there—in fact they proved such a success that a few, who were expert in handling them, made a business of contracting to raise chickens by their use up to a time past brooding age. This type of brooders have now been proved practical in all parts of the country.

Some people are fairly successful in raising young chickens in fireless brooders, and if carefully tended they may be raised this way in small flocks. The objection to this kind of contrivances, however, is that the attendant must be constantly "on the job" or disaster follows, and, if a number are in use, the labor and time items are serious ones to be considered.

For the poultryman desiring to raise one hundred to one hundred and twenty-five chickens, the portable brooders or hovers, which can be set anywhere on the floor of a house or shed, will prove practical. Each make of these brooders will have manufacturers' directions which should be followed.

For the poultryman in a larger way such units could be multiplied, or a larger brooder could be used to care for from five hundred to fifteen hundred chickens.

Let us now look at the requirements for successful brooding. Under hens, when the chicks feel chilled they huddle under the hen's body, and are warmed by direct contact with the same. Not only so, but the hen's intelligence is used to prevent the chicks getting chilled, because she watches over them and calls them to her to be huddled.

Under artificial brooding, there are three main requisites; first, that a warm place be furnished where the chicks can hover, second that the air in or under this hover be kept pure and wholesome, and third, that the chicks have no opportunity or excuse for crowding.

When outdoor brooders are used, these provisions are fairly cared for if the brooders are run at the right temperature. When run at too low a temperature the chicks will crowd; and, if run at too high a temperature the chicks may not be able to get away from the heat; and, if overheated the probabilities are they will never entirely regain the stamina lost thereby, and they will always be backward in growth as a result of this overheating. These are some of the reasons in favor of brooding with the movable hovers.

The portable hovers can be set on the floor of a house or shed, and they should be heated up some hours before the chicks are placed under them. Some straw or other litter should be spread on the floor and under the hover, and the chicks should be placed under the hovers to become hover-broke.

Whatever style of brooder is used, it should be started up and be got in readiness several hours before the arrival of the chicks, in order that the flock of little birds will not be met with a chilly reception when they appear on the scenes.

Preparation is a sure preventative of several causes of trouble.

In trimming brooder lamps, either the wick should be turned up far enough to slightly snip the corners with a pair of shears, or the charred wick should be rubbed or wiped off to slightly rounded corners. A wick trimmed in this manner will not smoke, even if turned up to a relatively high flame. The least suspicion of a pointed flame at the side of the wick is likely to cause smoking and consequent danger from over-heated lamp founts.

About the last act of the chickens before they emerge from the shell, in the hatching process, is the absorbing of the yolk. This yolk takes thirty-six hours or more to digest, and the chickens not only do not need anything to eat or drink under thirty-six hours, from the time they are excluded, but it is generally harmful to feed them anything sooner than this. Advantage is taken of this provision in nature to ship "day old chicks" at this time. They can, at this stage, be shipped with perfect safety and with no ill effects a distance of one thousand miles or more.

The "day old chick" business has grown tremendously in the past three or four years. There seems to be less opportunity for disputes or dissatisfaction in the purchase of day old chicks, than in the purchase of eggs for hatching. On the arrival of the chicks, the purchaser can see and count what he has received; and on the other hand, the seller is saved the annoyance and dissatisfaction of claims and disputes from patrons, whose troubles are caused in a great many cases through the fault of others in the handling of the eggs in shipment; or, in a great many other cases, by the carelessness or ignorance of the purchaser during incubation.

When the chicks arrive, take a coil of wire, of one inch or half inch mesh poultry netting, one foot or more wide. Nail or staple one end to a standard, with a footing on it so it will not topple over. Have another standard on which to place the roll or coil of wire. Now place this wire around the hover, in such a manner that the chicks cannot get away from the hover more than a few inches, until they know or learn where the place is where they can get warmed up.

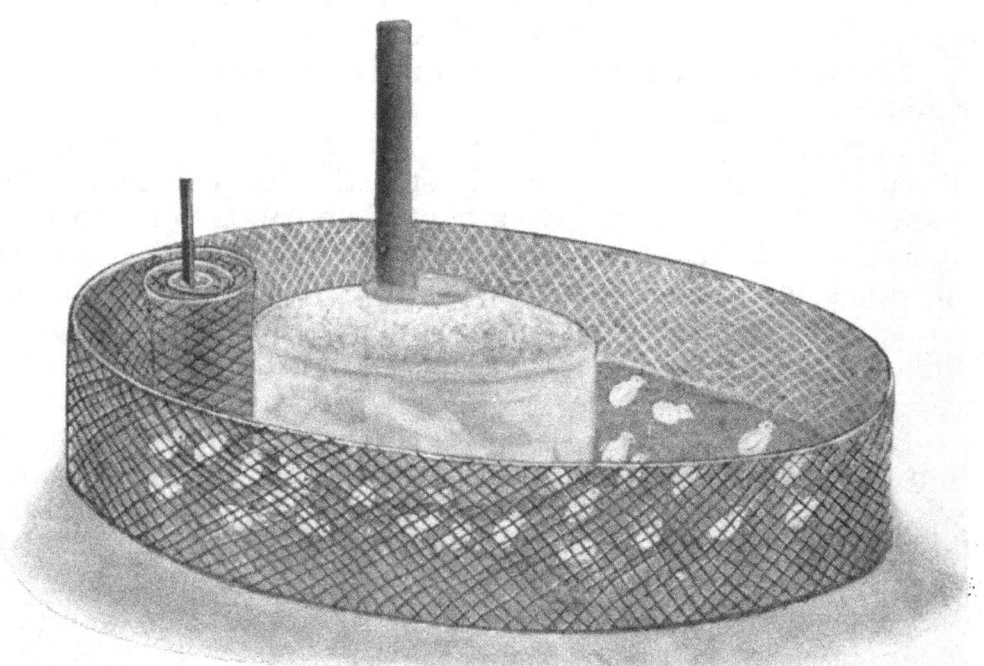

Enclosure for Young Chicks Described on This Page

When the chicks are hover-broke, loosen or unwind some of the coil of wire, so as to make a yard all the way around the hover, say three or four feet away; and give them their first food of fine grit of some kind. Sprinkle a little steel cut oatmeal (sometimes called pinhead oatmeal) around for them to pick up, and you will be surprised to see how soon they learn to eat, if you have no experience in this line.

Now provide water for them to drink, but use care as to how you provide it. Do not leave it where chicks can get into it, or step in it all the time; for, if you do, they will foul the water and get chilled by getting wet. A small automatic fountain can be used, and a few pebbles or stones can be placed at the drinking fount of same, for two or three days, to keep their feet from going in the water. This will also insure against some of the chicks being drowned, for it does not take much water to drown chicks a day or two old.

If you have no automatic fountain, you can take an empty tomato can, or a similar sized can, and find a saucer or other shallow vessel a little larger in diameter than the outside of the can. Punch two holes near the open end of the can, and at a depth so that when inverted, and placed in the saucer,

Gallon Tomato Can Drinking Fountain

the holes will be below the top of the same. You can now fill the tomato can with water, place the saucer over it and invert, and you will have a fine drinking fountain. If you place a weight on the top of the can, it will make it safer from danger of upsetting. The same style of fountain can be made of larger vessels for grown fowls, and will prove far better than open vessels, although not so convenient as regular automatic fountains, which can be purchased at very reasonable prices.

After the chicks get started eating and drinking, they can be fed on rolled oats or steel cut oatmeal, commercial chick feed and broken rice, or whatever other feeds in a fine state, and in variety, can be procured locally. Boiled rice is relished for a change, but must be fed with the moisture nearly all out, so it will not be mushy. This boiled rice helps out also in case of bowel trouble. Care must be used, however, not to leave any around to get sour.

You will also need charcoal, grit, and oyster shells.

Little chicks can be fed on boards with a strip of lath nailed around the edges, to prevent the feed from being scattered. If a board, about six inches high, is nailed upright in the center, and an opening is made for a handle, this kind of feeding trough will be very convenient, even for grown fowls, to feed a moistened mash. A feeding board, made in this manner, also protects the mash from being trampled on.

Growing chicks must be fed liberally if they are expected to mature early. A stunted chicken will never mature properly, or obtain its fullest possible growth, any more than will a stunted calf or colt.

If fed on the right kind of food, the writer does not think it possible to overfeed chicks after two weeks old. However full their crops, they are always ready for a taste of something new, and they should be given plenty of opportunities for gratifying this taste. We are considering now only chicks which are fed regularly—not chicks kept in an alternate condition of starvation and plenty.

In feeding chicks, we should not lose sight of the fact that we are manufacturing "egg machines," and chicks which never have their crops distended will develop small crops proportionately; hence, will have smaller fuel boxes for use in manufacturing eggs later. Growing chicks can be fed an extra meal late at night, if a good artificial light is used.

The writer has raised Leghorn chicks now for several years, and has always collected the first pullet eggs in from one to three weeks under five months from hatching. He has also seen chickens hatched from his own eggs, at the same time, and in the same incubators, which, for three or four seasons running, did not lay an egg before from seven to seven and a half months from hatching. It all depends on the care.

and liberal feeding, which the chicks receive while obtaining their growth.

Nothing is so dangerous to growing chicks as crowding, and the greatest care must be used to see that they do not get to feel chilled. This care must be taken even after they have started to roost, if the weather suddenly gets cold. In such cases the heat should be started up again in the hovers if possible.

The greatest danger in crowding is when chicks get into corners; but they can be taught, if they are started right, to always go to the hover.

After the chicks are hover-broke and are four or five days old, they will, if of the active breeds like Leghorns, begin to jump over the wire enclosure previously described, if such an

Feeding Board Described on Page 40
(With Supports to Raise Above the Litter for Grown Fowls)

enclosure is used. When they do this, the "yard" can be extended by unrolling more wire, and leaving an opening where the ends come. This opening can be made at the side of the house nearest the light to make its finding easy. The chicks will learn at once that they can get out all around the house or shed. Litter should be all over the floor. At dusk they should be watched very carefully for a day or two, as some of them, maybe only two or three, will fail to find the opening to the hover.

You will be surprised and amused to see how hard they will work to find this opening, and after a day or two you will have no trouble. When the chicks are all in their yard, or after they are trained, you can close the opening to this yard at your last visit for the night; and place feed and water inside the yard ready for morning.

The floor inside this yard should always be kept covered with litter, and the yard can be narrowed down to within a foot or two all around the hover at night, till the chicks are a few days old.

The hover should be run at a good temperature, and this temperature can be judged as well by the action of the chicks as by the thermometer reading. Do not make the mistake that some people do of trying to have the chicks furnish the heat to keep the hover warm, from a mistaken sense of economy and saving in fuel. It is safe to err on the side of high rather than low temperature, with this kind of hover, because the chicks can get away to where it suits their comfort to be.

Too much emphasis cannot be laid on the danger from **crowding**. If the temperature gets too low, and the chicks crowd badly, several will be smothered and serious loss will result; but the loss from dead chicks does not begin to tell the story. Three-fourths of the lot of chicks may be practically ruined by one night's crowding. Many chicks may be crippled, and it will be found that once a chicken has been half smothered in crowding it never gets over it—it may live but seems to have lost stamina.

Relatively high temperatures under the hovers obviates this danger, and herein lies the success of the large brooders mentioned, because the heat keeps the chicks at a distance and scattered. The same thing happens with the portable hovers, if run at high enough temperatures. Ninety to a hundred degrees under the hover, when no chicks are under it creating heat, will keep the chicks scattered around the outside on the litter.

Care must be exercised to keep the floor and litter dry. The litter under the hover should be cleaned out and renewed every three or four days.

If the hover is run "right" the chicks after a day or two will remain, when they "retire," all around the hover on the litter with their heads facing outward; so that if there are a lot of them they will appear as a mat all around the hover— each chick close against the others, but not crowded. As the night gets cooler, you will find them gradually moving to the warmth under the hover.

The advantages of these hovers, run in this way, over outdoor brooders are that, when it gets daylight, at from three to four a. m., the chicks do not have to wait for the attendant before they can obtain plenty of fresh air, exercise, feed, and water, but are ready for business as soon as they can see.

They can also be made perfectly safe from rats, cats, or other vermin, and they need not be let outside at all, if properly fed, until they are four or five weeks old.

Green stuff must be provided for the chicks regularly. A little chopped cabbage, some lettuce, weeds, lawn clippings, sprouted oats, etc., will provide this satisfactorily.

Keep the chicks growing, and feed them beef scraps or cut bone, after they are three or four days old.

If you have a garden, you will provide exercise, food, and excitement for the chicks, and great amusement for yourself, if you dig up a few worms occasionally and let the chicks have them. If you have an enclosed yard, as suggested, around the hover, you will have a regular circus and race track exhibition, both inside and outside the fence.

What Breed?

One of the first puzzles a novice in poultry keeping will face, will be what breed to choose for his particular purpose.

If the production of table poultry is the object, the choice will likely fall on the heavy Asiatic fowls, or a compromise of the middle weight breeds. If the object is a combination of eggs and table poultry the choice will likely be one of the middle weight breeds. If the primary object is the production of eggs in large numbers, the Mediterranean classes will furnish opportunities for choice.

The fancy of the poultryman will often decide what breed to choose—he may have a leaning to some particular breed.

Whatever breed is chosen, as much depends on the management of the flock as on the breed. A good poultryman will succeed with comparatively poor stock, while a poor poultryman will fail with the best stock obtainable.

There are several reasons why only thoroughbreds of some variety should be kept; and also why only one variety should be kept on the plant—in other words "one farm one breed." The owner will take much pride in having uniform flocks. This pride will affect his success in various ways. There is no pleasure in showing one's flocks to friends, neighbors or visitors if they are a lot of scrubs or mongrels. When a breed of thoroughbreds is kept the opportunity will come to sell eggs for hatching (or day old chicks if the plant is a large one), which will add considerably to the possible profits.

However clever the poultryman, and however great his success in handling his flocks, if he has ambitions, he can never attain the pinnacle in the poultry business unless his stock is thoroughbred.

Another reason which will make it desirable to keep thoroughbreds, of only one variety or breed, is that each variety or breed needs different treatment and feeding for results.

Of still greater moment to the beginner, or struggler after success, is the extra work entailed in keeping flocks separate, when more than one breed is kept, and the constant watchfulness necessary, during the breeding season, to insure against a mixture of the varieties of fowls.

Meat Production a By-Product.

To the poultryman who has chosen his breed with a view to profitable egg production, the sale of chickens or fowls for market will be a by-product proposition.

Young cockerels and the cull pullets may be sold either as squab broilers, or, from the time when about three months old or more, as spring chickens.

A hen will not continue to lay indefinitely, and the profitable period of egg laying will not extend much over three years from the time the birds are matured.

After a poultryman has got his start, it will therefore be necessary to replace some of his stock with young birds annually; and he will thus have some of the older birds to send to market each year.

Oil or other fuel, litter, and feed will cost from between eighteen and twenty-five cents per chicken to raise them to the age of there months.

After this age, it will cost more to feed them. They then develop faster and use more feed. An allowance may be made of from ten to twelve cents per month to feed the pullets for the next three months, by which time (when they are six months old) they should be producing eggs if properly fed and matured.

We thus have an approximate cost of between forty-eight cents and fifty-five cents for food only, to produce a pullet to laying age.

If day old chicks are purchased, the cost for good stock will be from twelve to twenty-five cents each.

If the chicks are hatched from one's own eggs, there will probably be a general average of about fifty per cent hatched from all the eggs set.

Hatches of fifty per cent of all the eggs set will be considered a conservative estimate by most poultrymen—many would be pleased to do that well.

If we figure the price of eggs (as table eggs) at thirty cents per dozen, we will, with a fifty per cent hatch, get birds at a cost slightly in excess of five cents each—making allowances for the cost of oil for incubators, or the feed of hatching hens.

We thus arrive at a probable cost of between sixty cents and eighty cents for matured pullets, if day old chicks are purchased; and between fifty-three cents and sixty cents if one's own eggs are used for hatching, based on the price of table eggs.

These figures do not allow for any losses from mortality through sickness or accident. Losses of this nature will vary from ten to twenty per cent—sometimes more—depending on the stamina of the flock, or on the intelligent care they receive.

Even at the lower probable cost, here given, of fifty-three cents (when chicks are hatched at home from one's own eggs, with whatever more mortality costs are to be added thereto) it will be readily seen that with the lighter breeds, such as Leghorns, which would only weigh about three pounds at this age, the poultryman will have to look to his egg production, rather than to the meat market, for his profits.

As a by-product, however, there is some satisfaction in the knowledge that after a period of two and one-half years, from this time, of profitable laying the pullets will, with a slight increase in their weight at three years of age, bring about as much on the market, if in good condition, as they will when first matured.

The poultryman will also have the satisfaction of knowing that the period of his greatest risks are about over at this time. Broilers die off faster than mature fowls from various causes.

When raising a flock of layers, the poultryman will get some relief in his expense account when the cockerels and the cull pullets are sold. Leghorn cockerels will weigh from one to one and a half pounds at ten weeks of age.

Whether the cockerels are sold, or not, they should be separated from the pullets when they are from six to seven weeks old. This separation will give the pullets a better opportunity for development. Leghorn cockerels mature early and are very precocious—we have had them crowing under six weeks of age. Both the pullets and the cockerels will develop faster when separated.

For the reasons given, the egg poultryman will get his eggs first, and will endeavor to sell his meat by-product for as much, after taking his profits, as he could have realized from his birds at maturity.

The figures in this chapter are, of course, only estimates. In actual practice better results may be obtained. Figuring and estimating to be safe must take account of risks and losses; and these probable losses and risks should be discounted before the start—then, if any of them are escaped, so much the better.

Yards and Exercise.

The space devoted to yards, or runs, for the poultry will depend to a great extent on the amount of space available. The poultryman on the city or suburban lot is necessarily limited in his yard space.

To the poultryman with unlimited ground, the amount of yard or run space will vary, according to his judgment and available funds for fencing, etc. The suburbanite need not be discouraged by his restriction in space, for, if a small plot is taken care of, and his flock has all the necessary food elements supplied to them, he very often will find his production average, or percentage, far exceeding that of the poultryman with unlimited space.

Small yards should be spaded up occasionally, and should be limed at intervals of a few months, to keep them in a safe and sanitary condition. This spading will be the more necessary according to the nature of the soil. Clay soils or soils of a close texture soon become foul, and trampled down, so as to make them impermeable to water; whereas open, sandy, or gravelly soils allow the impurities to sink down into the sub-soil with every heavy rainfall.

In its effect on egg production, runs which are very large may cut down this production by excess of exercise. Under certain circumstances, the food picked up may be mostly used up in muscular effort.

A flock in good laying condition, and producing heavily, which had been confined in moderate sized yards, and then allowed to run over a wide stretch of ground all day, would almost surely drop off in egg yield in a day or two. The same flock if allowed only an hour or so, at or near dusk, so that they will return of their own accord, over the same ground, would very likely increase their egg yield by reason of this liberty.

The records given by the writer were made in yards without a blade of green stuff growing therein, and the fowls were not outside of these yards over thirty minutes in three hundred and sixty-five days.

Too much exercise reduces flesh and makes muscle—so much so that some butchers object to, and will refuse to purchase, Leghorn hens for marketing. Butchers have a basis for this objection in many cases. Fowls of the variety in question when half fed, or when forced to work hard for every bit of food they procure, will be tough eating.

A Leghorn hen at the age of three years, that has been properly fed for egg production will, if in good condition for marketing, prove as good eating and as tender as any other variety of matured fowls at any age. We have eaten Leghorn hens from our yards when over four years old, which would be boiled tender in an hour and a quarter, and would separate from the bones when boiled for an hour and a half.

Breeds like the Leghorns will make opportunities for themselves to take plenty of exercise, without providing large yards.

Leghorn roosters will very likely be tough at any age after six or seven months, unless confined in a small coup or pen to reduce exercise, and fed liberally for a week or two before being dressed. If this is done, however, they will cook tender and prove good eating.

In a few words, looking at this matter from a producing standpoint, if laying hens are expected to "pick up their living," in order to economize on the feed bill, the economy in feed will be **much more** than offset by the domestic economy of the hen in a reduced egg yield.

That Satisfied and "Comfy" Feeling Which Helps to Fill the Egg Pail (See Page 10)

CHAPTER IV

Production and Care of Eggs

Spring and Summer Eggs.

Spring and early summer are the natural seasons for egg laying. Even in a wild or natural state this truth holds good. At these seasons the temperature is such that there is no great amount of food needed to keep up the bodily heat, and the necessary elements for egg producing are in great abundance.

Green stuff is plentiful, animal food is easily picked up in the shape of worms, insects, bugs, etc.; and numerous elements are detected by the sharp eyes of the hens, either on the bare ground or when brought to light by scratching.

Any hen that will ever produce eggs will do so at this time of the year, and the efforts of poultrymen for a great many years have been bent to simulate these conditions as much as possible, in order to increase production; for, if the hens produce under these summer conditions, the reasoning seems sound that the nearer the approach to these conditions, at other seasons, the better the yield.

This reasoning has proved out in actual practice, and production has steadily increased in proportion as these conditions have been furnished.

From the wild jungle fowl laying twenty to thirty eggs annually to the general average for the United States of about sixty is quite a step, and the general average for well cared for farm flocks will run from sixty to eighty eggs per hen annually.

Of course, if the food elements are supplied in abundance, production will be greatly increased, even in spring and summer.

In the latter part of spring and early summer, broodiness of the hens cuts down the output considerably, unless this matter is watched.

This condition also spoils a great many eggs, because the eggs are not gathered often enough, except by regular poultrymen, and broody hens are allowed to sit on the eggs all day or longer.

The Asiatic and the heavy American breeds are especially troublesome in this respect. Where only a few fowls are to be cared for, one does not notice the trouble and bother of these broody hens very much; but to the poultryman caring for from seventy-five to one thousand hens, or more, it becomes quite a problem.

The lighter Mediterranean breeds do not give so much trouble by broodiness, although even the Leghorns are sometimes persistent sitters.

Taking broody hens off the nests the first time, or so, when they stay on the nests at night, and placing them in a slatted coup with a slatted bottom, where they can get food and water, and be in sight of the other fowls, will break up most of the lighter breeds in a day or two, when they can be let out to start getting ready to lay again. The slatted bottom on a coup allows a circulation of air to cool their bodies, and thus helps to eradicate the sitting fever.

As the summer draws to an end, the egg production, if the hens are not fed properly, gradually dwindles down to almost nothing.

There are several reasons for this: the discomfort from the heat, lack of shade, dead air in the poultry houses, because no provisions have been made to have the air circulate, scarcity of good drinking water, and an irregular supply of water.

It is false economy in every way to allow hens to be without water for one minute of their **business** hours. The composition of an egg is nearly two-thirds water, and to withhold this, the cheapest element in production, is indeed short sightedness. Hens which are laying well, if allowed to really suffer from thirst for half a day or more, will very likely drop off in egg yield at once, and may not get back in form again for a week or more.

Fall and Winter Eggs

In late summer, and early fall, the hens will shed their feathers, or go through the moult. This process is the most trying of any through which the hens have to pass, and under unfavorable conditions will prove a severe drain on their constitutions and vitality. The losses from debility, at this critical time, will exceed those at any other season of the year.

It can be noticed that, even in vigorous stocks of young chickens, perceptible and rapid growth does not seem to start until the birds complete their first coat of feathers, to replace the downy feathers they have when hatched.

Liberal feeding should be resorted to during the moult, and, as the natural supply of animal food in the shape of worms, bugs, etc., is much less than early in the season, this should be supplemented by a more liberal allowance of animal food in the shape of beef scraps, green cut bone, ground fish bone, etc.

Farm poultry generally quit laying at this time of the year, and **lay off** until spring. Fowls which are liberally fed, on a well balanced ration, will continue to lay while shedding their feathers; but nearly every hen will stop laying for a longer or shorter period, while growing her new feathers.

Early pullets, if properly matured, will have commenced laying by this time, and will help to keep up the regular supply of eggs.

If the hens get completely through the moult, they will start to lay again and will lay more or less throughout the winter under proper care, and with **extra liberal** feeding.

Summer conditions (or spring and early summer conditions) are what have been preached at us by the poultry press for many years. These conditions have been stated as variety of feed, green feed in abundance and in a succulent state, and animal food. We cannot, however, supply the climatic conditions without heating arrangements, and nearly all attempts at keeping matured fowls in heated houses have proved failures, on account of their debilitating effects.

We must overcome this difference in temperature by extra liberal feeding, if we are going to obtain results in eggs. It is going to cost money to keep a flock of hens over winter; and if we feed only enough to keep a hen in good condition, we are out that amount of money. Many flocks remain on the verge of laying all winter but do not start to lay until spring, because of the short sighted policy of their owners in furnishing just enough feed to keep them in good condition, and withholding the extra feed required which would produce the eggs to pay their board, and leave a surplus as a profit.

Winter eggs cost money—they cost, in most cases, far more than they will sell for—even if sold as high as seventy-five cents per dozen. Even when obtained in winter, in liberal quantity, eggs will pay a better profit at thirty cents per dozen in spring than at fifty cents per dozen in winter, because it takes much less feed to maintain the fowls in spring, and this leaves a surplus of food and· energy to produce more eggs.

The quantity of feed required for good production is astounding to most people; and these quantities can be judged by the statistics shown later on in this book. (Pages 56 and 79

However, if hens do not pay any of their feed bills in winter, the profits of spring and summer will mostly disappear in making this loss good. We must, therefore, strive to get eggs in winter, even if we have to sell them at fancy prices below cost of production.

Care of Eggs

Fresh laid eggs are a delicacy, but to retain this delicacy they must be handled and cared for in a delicate way. The shell of an egg is porous, so porous that, during the process of hatching, air is supplied to the growing chicken through the shell.

If an egg is kept in an unsanitary condition or near strong odors, it will be affected thereby. If eggs are kept on musty hay, or packed in a case where a little musty hay is used as part of the packing, they will become so musty as to be inedible, although otherwise in good fresh condition. Nest eggs are made, and are offered for sale, which contain carbolic acid and other disinfectants, on the theory that they keep away lice from the nests. It will be found that where these are used, when the hens begin to get broody, the eggs will be flavored by such nest eggs.

Eggs where roosters run with the flock will also be fertilized, and therefore contain potential life—only needing temperatures of eighty-five degrees or ninety degrees, and above, to start this potential life into growth.

The process of successful incubation only requires temperatures between one hundred degrees and one hundred five degrees without much variation, for long periods, above or below this range of temperature, and certain conditions of pure air and moisture. There is nothing offensive about a hatching egg—the offensiveness comes in as a result of a start in incubation, and a subsequent death of the germ.

Eggs, therefore, must be kept at cool temperatures during the whole period between the nest and the table, and in sanitary and odorless places.

Some eggs may be laid containing blood spots, and, while these eggs may be perfectly good, they should not be packed to be sold with high class eggs. To obviate the possibility of this happening, eggs should be "candled" the same day as laid, and eggs showing blood spots should be culled out.

The cause of blood spots in eggs seems to be an open question, which the writer does not presume to settle. He has found in his experience, however, that a frequent occurrence of spots in eggs happened when there was an **excess** of roosters in the pen; and that the trouble stopped almost immediately when the number of roosters was reduced.

Another mooted question is whether hens or pullets will lay as well or better with roosters in the flock, than they will if roosters are not in evidence. The more natural condition seems to be to have the roosters run with the flock at all seasons.

The author is of the opinion that hens will lay better, and be more contented, with roosters running with them than they will be without them. He did not think it made very much difference in the egg yield until one season, being short of roosters, he put them all in one pen—separated from the next pen only by a wire netting. In the course of about two weeks the eggs from the roosterless pen began to be smaller, and several eggs were gathered too small for marketing. Being at a loss to account for this condition, he reversed matters by placing the roosters in this pen, and in a few days the trouble stopped. It was not long, however, until the other pen developed the same condition; and the results were so peculiarly interesting, that conditions were reversed once or twice more to make the test convincing.

To reach the market, and obtain the best price, eggs should be packed neatly and attractively. All the eggs should be clean and, for the best price, uniform in color and size. All misshapen eggs should be culled out.

"Ask the Birds, Their Judgment Is Good"

This suggestion, to "Ask the birds, their judgment is good," has appeared in several parts of the text of this little book, and will bear reiteration. A regard, or disregard, of the idea embodied in these words, will always mark the difference between the successful and unsuccessful poultryman.

The fancier, of course, will, by the very fact of his being a fancier, take pains to maintain his birds in comfort—he is in the business for the love of the game.

The poultryman who expects returns in eggs, in sufficient number to return him a profit, however, must be in touch with the needs of his flock. The question has been tritely put, "Why keep chickens? Let the chickens keep you!"

Another way of stating the fact that a poultryman must know his business, in order to draw dividends, was used by two men in discussing the poultry business—one of them remarked, "There is money in the chicken business," to which remark the other replied, "Yes, I know there is; I have put it there."

Now, unless one has an unlimited bank account, this idea, in money matters, of "putting it there" must soon come to an end; and it does come to an end for many who engage in this business of poultry keeping.

Rosy outlooks are presented of what may be expected from this business, and mathematical calculations are made of the increase in flocks, or the production in eggs, which cause many to enter a business of which they know nothing, and for which they have not the necessary qualifications.

It is true in all lines of business that "business is business," and the poultry business is no exception.

To be successful, "details" (that horrid word which so many detest) must be attended to, as a regular performance; and for this attendance many are not fitted by nature.

The business of producing eggs and poultry will not be overdone, if for no other reason than that it takes men of special qualifications to stay in the game successfully. There is no royal road to success in poultry keeping. All that can be learned from books on this subject are certain fundamental truths; but no person can give another the detail knowledge necessary to carry him safely through all his difficulties. The poultryman must have a large share of initiative qualities for his own guidance.

The action of the birds will always indicate, to the one possessing these qualifications, the needs to be met or furnished for the well being and comfort of his flock. Whether he puts it in so many words or not, even to himself, he knows enough to "Ask the birds, their judgment is good."

Production Under Present Methods.

Under the methods in present use, egg production has its periods of famine and plenty. In spring and early summer hens produce well, in late summer and early fall the yield falls off, and in fall and winter we have to rely on early hatched pullets to supply most of the eggs. Pullets hatched too early are prone to stop laying, and moult again when winter sets in, and not return to laying again before late winter or spring. The hens, if they moult well, and get completely over the moult before the advent of cold weather, will lay well in early winter, but will gradually drop off in egg yield after the weather gets colder.

Liberal feeding will produce good results, if the rations are fairly well balanced. The writer always succeeded in getting winter eggs, even when the weather reached low temperatures for long periods.

It was this fact that led to further research as to why, if some of the hens and pullets laid, more of them did not do so under his efforts to supply "summer conditions."

Every item of feed had to be purchased, as his flocks were continually kept yarded up.

To give an idea of the liberal feeding, and the resulting production, the following tables are presented for his flock for 1913. In this lot were ninety-six one and a half year, and two and a half year old hens, at the beginning of the year 1913, and forty pullets hatched May 3rd, 1912.

The record of the exact dates when some of these hens were sold, or dropped out of the race, is not available, so that a correct percentage cannot be figured out for this year. They were sold off down to ninety-five during the spring of 1913, in order to make room for day old chicks arriving on April 28th.

At the end of 1913 we had ninety-five hens—some one and a half years, some two and a half years, and a few three and a half years old; and sixty-five pullets hatched April 28th, 1913.

Following on this page are the exact figures for feed purchased. This feed was used to raise the young stock as well as being fed for egg production.

The writer will cheerfully admit some waste in feeding; but, as an explanation, of this waste, he had to leave home **every** morning (except Sunday) before seven a. m., and did not return home again before seven-thirty p. m., and often not before ten p. m. Under these circumstances, in order to make the best production, he decided to provide an excess rather than a shortage of feed—hence some waste.

The detail records of the egg production and comments on them will appear in the following pages:

Feed Purchased in the Year 1913

Date		Weight in Pounds	Variety	Total
January	1	100	Scratch Feed	$ 1.75
"	4	100	Scratch Feed	1.85
"	8	100	Mash	2.50
"	8	100	Scratch Feed	1.85
"	13	100	Mash	2.15
"	13	200	Scratch Feed	3.50
"	16	100	1 bbl. Cabbage	1.75
"	16	64	2 bus. Oats	.80
"	20	100	Scratch Feed	1.75
"	22	100	Scratch Feed	1.69
"	23	55	Beef Scraps	1.78
"	29	100	Scratch Feed	1.69
Total.		1,219		$23.06
February	3	200	Scratch Feed	$ 3.40
"	3	100	Mash	1.89
"	10	200	Scratch Feed	3.38
"	10	100	Oyster Shells	.80
"	13	100	1 bbl. Cabbage	1.50
"	17	100	Scratch Feed	1.73
"	21	100	Scratch Feed	1.72
"	25	100	Scratch Feed	1.85
Total.		1,000		$16.27

Date		Weight in Pounds	Variety	Total
March	3	50	Beef Scraps	$ 1.63
"	3	200	Scratch Feed	3.30
"	4	100	1 bbl. Cabbage	1.50
"	7	100	Screenings	1.39
"	8	100	Mash	2.50
"	8		1 bale Straw	.35
"	8	64	2 bus. Oats	.80
"	12	100	Mash	1.79
"	12	100	Scratch Feed	1.85
"	12	100	Screenings	1.39
"	13	100	Scratch Feed	1.85
"	19	100	Scratch Feed	1.85
"	21	100	Charcoal	2.25
"	22	100	Scratch Feed	1.85
"	26	100	Scratch Feed	1.85
"	27	100	1 bbl. Cabbage	1.40
"	28	100	Scratch Feed	1.85
"	31	100	Scratch Feed	1.85
Total.		1,714		$31.25
April	1	50	Beef Scraps	1.63
"	1	200	Scratch Feed	3.70
"	1	100	Screenings	1.39
"	10	100	Mash	2.25
"	10	100	Scratch Feed	1.85
"	12	100	Scratch Feed	1.85
"	17	100	Scratch Feed	1.85
"	22	200	Scratch Feed	3.70
"	22	100	Mash	2.25
"	23	50	Chick Feed	1.10
"	24	50	Chick Feed	1.10
"	25		1 bale Straw	.40
"	25		2 gal. Oil	.33
"	25	10	Fine Grit	.09
"	25	4	Oatmeal	.20
"	26	5	Rice	.30
"	26	5	Oatmeal	.25
Total.		1,174		$24.24
May	1	100	Scratch Feed	$ 1.85
"	1	50	Beef Scraps	1.63
"	1	50	Oyster Shells	.40
"	1	50	Grit	.40
"	3	64	2 bus. Oats	.80
"	6	100	Mash	2.25
"	6	200	Scratch Feed	3.70
"	9	200	Scratch Feed	3.40
"	22	100	Mash	2.25
"	29	100	Scratch Feed	1.85
Total		1,014		$18.53

Date		Weight in Pounds	Variety	Total
June	2	250	Scratch Feed	4.25
"	2	50	Chick Feed	1.25
"	2	50	Beef Scraps	1.63
"	2	100	Mash	2.25
"	2	50	Alfalfa Meal	1.25
"	6	50	Oyster Shells	.40
"	6	50	Grit	.40
"	6	100	Mash	2.25
"	9	128	4 bus. Oats	1.80
"	11	150	Scratch Feed	2.49
"	11	50	Screenings	.60
"	13	50	Chick Feed	1.25
"	14	100	Cracked Corn	1.39
"	18	100	Scratch Feed	1.85
"	19	200	Toasted Corn Flakes	2.00
"	24	100	Scratch Feed	1.85
"	25	200	Scratch Feed	3.40
'	30	10	Grit	.09
Total.		1,788		$30.40
July	1	50	Beef Scrap	$ 1.63
"	1	50	Chick Feed	1.25
"	3	160	Mixed Meal	2.00
"	5	100	Mash	2.25
"	7	100	Scratch Feed	1.85
"	10	100	Scratch Feed	1.85
"	14	300	Scratch Feed	5.55
"	16	50	Beef Scraps	1.63
"	16	50	Oyster Shells	.40
"	16	100	Mash	2.25
"	19	50	Mixed Feed	.85
"	21	250	Scratch Feed	3.95
"	26	250	Mixed Meal	3.13
Total.		1,610		$28.59
August	1	100	Mash	$ 2.25
"	1	100	Scratch Feed	1.85
"	1	128	4 bus. Oats	1.84
"	2	50	Beef Scraps	1.63
"	12	200	Scratch Feed	3.70
"	12	100	Grit	.80
"	18	100	Mash	2.25
"	18	100	Scratch Feed	1.85
"	21	200	Scratch Feed	3.70
"	29	100	Scratch Feed	1.95
Total.		1,178		$21.82

Date	Weight in Pounds	Variety	Total
September 3	50	Beef Scraps	$1.63
" 3	100	Scratch Feed	1.95
" 3	100	Oyster Shells	.80
" 3	100	Mash	2.25
" 7	300	Scratch Feed	5.40
" 7	100	Mash	2.25
" 10	200	Scratch Feed	3.60
" 18	50	Alfalfa Meal	1.00
" 18	50	Beef Scraps	1.63
" 18	100	Mash	2.25
" 27	10	Oil Meal	1.00
" 27	100	Scratch Feed	2.10
Total	**1,260**		**$25.86**
October 1	200	Mash	4.50
" 1	200	Scratch Feed	4.20
" 3	50	Beef Scraps	1.63
" 4	126	Yellow Turnips	1.73
" 9	200	Scratch Feed	4.20
" 13	128	4 bus. Oats	1.92
" 13	100	Scratch Feed	1.90
" 23	100	Mash	2.25
" 25	100	Scratch Feed	1.90
" 30	100	Scratch Feed	1.90
Total	**1,304**		**$26.13**
November 3	50	Beef Scraps	1.63
" 3	200	Mash	4.50
" 3	100	Scratch Feed	2.00
" 5	100	Scratch Feed	1.85
" 12	100	Scratch Feed	2.00
" 13	100	Scratch Feed	1.85
" 18	50	Mixed Meal	.80
" 19	100	Scratch Feed	1.85
" 22	100	Scratch Feed	1.85
" 25	100	Scratch Feed	2.00
" 26	100	Scratch Feed	1.85
" 29	100	Mash	1.35
Total	**1,200**		**$23.53**

Date	Weight in Pounds	Variety	Total
December 1	100	Scratch Feed	2.00
" 1	50	Beef Scraps	1.63
" 3	100	Mash	2.25
" 4	500	Scratch Feed	9.00
" 6	64	2 bus. Oats	1.00
" 6	200	Mash	2.73
" 9		1 bale Straw	.45
" 13	100	Grit	.80
" 13	100	Oyster Shells	.80
" 15	500	Scratch Feed	9.00
" 30	100	Yellow Turnips	1.65
" 31	100	Mash	1.35
" 31	64	2 bus. Oats	1.00

Total. 1,978 $33.66

RECAPITULATION FOR 1913.

Month	Weight of Feed Purchased	Cost of Feed	Number of Eggs	Value of Eggs
January	1,219	$ 23.06	805	$ 29.37
February	1,000	16.27	1,095	33.66
March	17,14	31.25	1,938	53.68
April	1,174	24.24	2,167	50.14
May	1,014	18.53	1,900	47.87
June	1,788	30.40	1,652	36.60
July	1,610	28.59	1,299	31.94
August	1,178	21.82	1,192	21.82
September	1,260	25.86	719	20.97
October	1,304	26.13	237	7.91
November	1,200	23.53	668	26.53
December	1,978	33.66	1,057	43.47
Totals	16,439	$303.34	14,729	$403.96

One hundred seventy-five eggs were sold for hatching for $11.50.

MEAT SOLD IN 1913

Months					
April	19.......... 2 Hens	6½ pounds	18c	$ 1.17
"	19.......... 8 "	25½ "	18c	4.59
"	26.......... 2 "	7 "	17c	1.19
"	28.......... 8 "	26 "	17c	4.42
May	26.......... 6 Roosters	24½ "	15c	3.67
July	26.......... 2 Cockerels ...	3¼ "	22c	.71
"	29..........15 " ...	22 "	21c	4.62
August	2.......... 1 " ...	1½ "	24c	.36
September	1.......... 1 " ...	2½ "	20c	.50
"	6.......... 4 " ...	10 "	20c	2.00
"	13.......... 3 " ...	8½ "	20c	1.70
"	20.......... 3 " ...	10½ "	20c	2.10
"	27.......... 4 " ...	12 "	20c	2.40
October	4.......... 3 " ...	9½ "	14c	1,33
"	11.......... 2 " ...	7½ "	15c	1.13
"	18.......... 1 " ...	3¾ "	15c	.56
"	25.......... 1 " ...	3¾ "	16c	.60
November	26.......... 2 " ...	9 "	12c	1.08
December	28.......... 3 Roosters	14¼ "	10c	1.43
Totals		207½ "		$35.56

As stated before, the record of all the hens sold and the dates of same are not available, some data having become lost. These records were kept for private use, never thinking they would be needed for publication.

Production Records for 1913

As will be seen in the previous pages, the straw purchased has been included in the feed cost.

All the oil purchased was not entered, as oil was used for other purposes and the account was not kept separate.

The feed, in addition to raising the young stock up to laying age, was also fed to the cockerels until disposed of.

If we accept the cost of eggs for hatching, the raising of young birds, and the keeping of cockerels until disposed of, as part of the necessary expense of egg production—which is offset in a more or less exact ratio by the proceeds of the sale of the old stock to replace which they are raised—we can see that it took 16,439 pounds of feed to produce 14,729 eggs, in the year 1913, by this flock.

The tables shown in this chapter give the daily egg production. The production of the older fowls is shown separately from that of the pullets from January 1st to April 21st. At this time the flocks had to be consolidated, to make room for day old chicks. The production is shown separately again each day from September 28th, when the pullets commenced laying, until the end of the year.

The reader will note the following comparisons of the laying of the birds, in their pullet year, with the laying by the adult fowls, in the fall and winter months.

1913	Number of Eggs Laid by Pullets	Number of Eggs Laid by Hens
January	588	217
February	558	537
October	130	107
November	560	108
December	810	247
Total for five months	2,646	1,216

There were over ninety birds among the adult fowls, and only an average of about fifty pullets. We can see by this comparison that the young birds have a decided **natural** advantage over the adult fowls, for fall and winter laying.

For the period from January 1st to April 21st, the pullets of the previous year and the old birds compare as follows:

1913	Pullets	Hens
January	588	217
February	558	537
March	690	1,248
April 1st to 21st	466	1,096
Totals for the period	2,302	3,098

This comparison shows the old hens as being able to produce heavily, as soon as spring conditions arrive. This is true, even with the handicap they had in January, when the pullets had a lead of 369 eggs, and a slight handicap in February of 21 eggs.

To recapitulate, we find as follows:

Weight of feed, 16,439 pounds; cost of feed, $303.34; number of eggs, 14,729; value of eggs, $403.96; meat sold, $35.56.

Allowing the meat sold to offset the feed used in raising the young stock, and the cost of eggs for hatching at prices for table eggs, we have a showing as an approximate cost of 24.71c per dozen for feed for 1,227 5-12 dozen eggs; and we find that it took 13.39 pounds of feed to produce a dozen eggs.

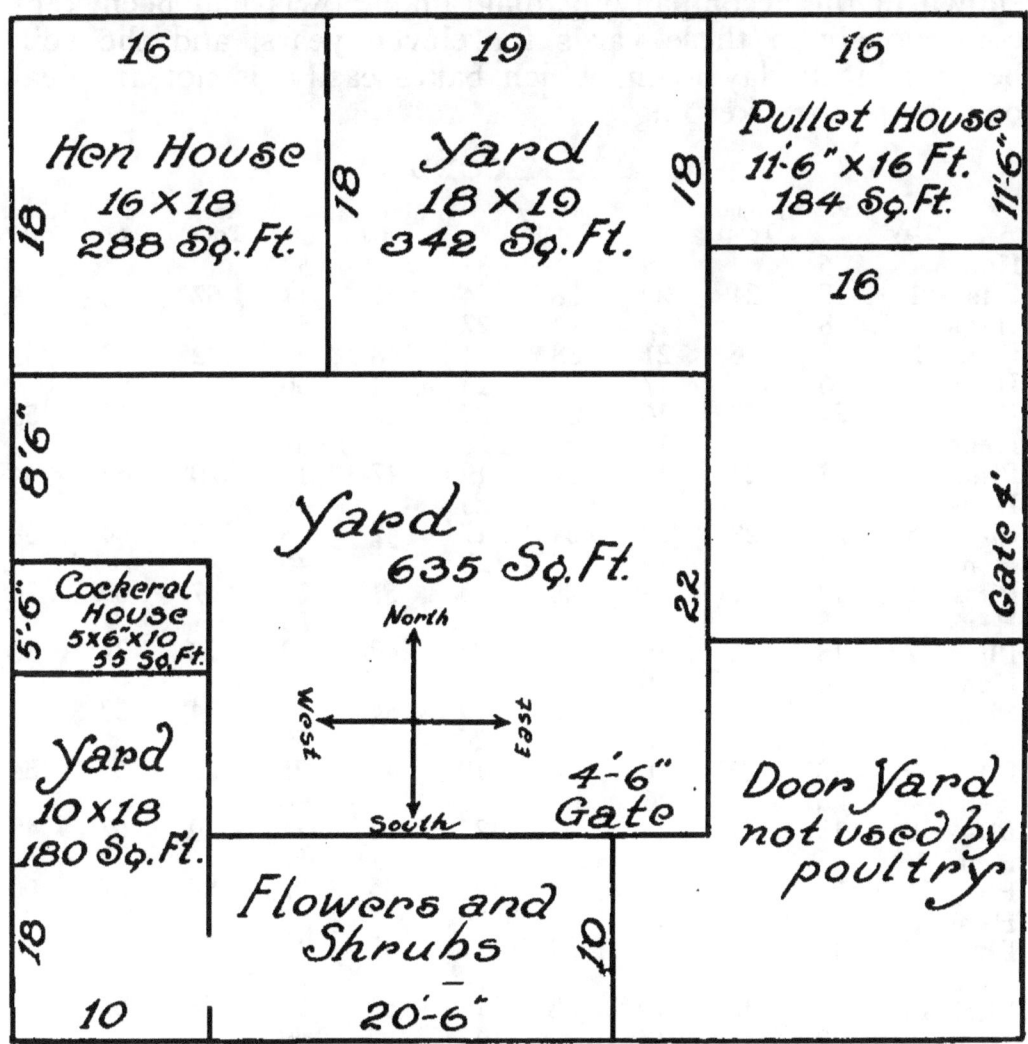

Diagram of Yards Used in These Experiments

The eggs were sold at an average price of 32.91c per dozen, this price being 8.20c per dozen above the cost of feed.

The reader is reminded that every item of feed for this flock had to be purchased, because there was no opportunity, at any time, for the birds to "pick up their living." They had in addition to the feed purchased, a few table leavings from a family of four, but these leavings do not count for much in a flock of this size.

The average cost of feed was $1.85 per hundred pounds for the year.

The space occupied consisted of yards and houses as shown in the accompanying diagram. Fowls had been kept continuously in these yards for eleven years, and the soil, being a black clay loam, which bakes easily, is not an ideal one for poultry keeping.

EGG RECORD

	Day	1913—January	Total	February	Total	March	Total	April	Total	May Total	June Total
Hens		6		8		37		53			
Plts.	1	18	24	20	28	24	61	14	67	62	65
Hens		8		17		27		51			
Plts.	2	18	26	21	38	21	48	21	72	72	65
Hens		6		17		25		50			
Plts.	3	21	27	16	33	22	47	25	75	63	55
Hens		6		14		28		51			
Plts.	4	17	23	17	31	19	47	19	70	63	64
Hens		7		12		25		49			
Plts.	5	21	28	22	34	17	42	23	72	64	58
Hens		4		14		28		46			
Plts.	6	18	22	14	28	23	51	21	67	69	56
Hens		8		9		26		55			
Plts.	7	18	26	19	28	22	48	23	78	71	60
Hens		8		16		35		57			
Plts.	8	19	27	16	32	19	54	17	74	58	60
Hens		7		10		26		50			
Plts.	9	18	25	20	30	19	45	25	75	55	54
Hens		6		13		38		48			
Plts.	10	15	21	18	31	25	63	21	69	67	48
Hens		5		12		29		55			
Plts.	11	22	27	21	33	25	54	27	82	57	66
Hens		9		18		36		53			
Plts.	12	19	28	20	38	21	57	18	71	54	48
Hens		4		19		33		54			
Plts.	13	14	18	19	38	17	50	24	78	67	52
Hens		7		13		42		49			
Plts.	14	20	27	19	32	24	66	28	77	59	57
Hens		5		21		37		58			
Plts.	15	16	21	17	38	24	61	23	81	53	50
Hens		5		19		40		47			
Plts.	16	15	20	22	41	22	62	23	70	63	50
Hens		7		22		39		50			
Plts.	17	17	24	20	42	20	59	30	80	63	55
Hens		5		18		45		60			
Plts.	18	19	24	22	40	24	69	21	81	53	52

Day	Type	January	Jan. Total	February	Feb. Total	March	Mar. Total	April	Apr. Total	May Total	June Total
	Hens	9		23		45		59			
19	Plts.	24	33	23	46	25	70	23	82	80	46
	Hens	10		21		46		51			
20	Plts.	12	22	21	42	25	71	21	72	53	57
	Hens	4		31		50		50			
21	Plts	21	25	23	54	24	74	19	69	59	46
	Hens	6		24		42		——			
22	Plts.	22	28	21	45	18	60	1,096	69	67	66
	Hens	7		25		52		466			
23	Plts.	14	21	24	49	24	76		68	66	47
	Hens	7		27		46					
24	Plts.	22	29	16	43	20	66		83	55	57
	Hens	7		28		52					
25	Plts.	22	29	23	51	26	78		70	62	50
	Hens	8		27		56					
26	Plts.	23	31	22	49	21	77		63	56	51
	Hens	5		35		56					
27	Plts.	22	27	23	58	20	76		67	58	50
	Hens	10		24		50					
28	Plts.	23	33	19	43	26	76		69	55	50
	Hens	10				57					
29	Plts.	15	25			23	80		65	66	66
	Hens	10				47					
30	Plts.	22	32			27	74		51	54	51
	Hens	11				53					
31	Plts.	21	32			23	76			56	
Tot. Hens		217		537		1,248					
Tot. Plts.		588		558		690					
Grand Total		805		1,095		1,938			2,167	1,900	1.652

Day	Type	July Total	Aug. Total	Sept. Total	Oct.	Oct. Total	Nov.	Nov. Total	Dec.	Dec. Total
	Hens				9		4		7	
1	Plts.	47	41	37	1	10	8	12	36	43
	Hens				8				12	
2	Plts.	51	42	31		8	6	6	27	39
	Hens				6		2		6	
3	Plts.	50	45	31	1	7	9	11	35	41
	Hens				5		2		11	
4	Plts.	42	41	30	1	6	10	12	32	43
	Hens				9		2		9	
5	Plts	34	43	30		9	10	12	29	38
	Hens				4		2		7	
6	Plts.	34	50	29	1	5	6	8	28	35
	Hens				6		1		10	
7	Plts.	28	47	40	1	7	12	13	28	38
	Hens				8		3		10	
8	Plts.	27	44	24	1	9	10	13	30	40

	Day	July Total	Aug. Total	Sept.	Sept. Total	Oct.	Oct. Total	Nov.	Nov. Total	Dec.	Dec. Total
Hens						2		1		7	
Plts.	9	36	43		30	1	3	12	13	28	35
Hens						5		1		10	
Plts.	10	26	41		35	3	8	12	13	25	35
Hens						4		3		8	
Plts.	11	40	38		33	4	8	9	12	31	39
Hens						1		2		10	
Plts.	12	34	37		25	1	2	10	12	27	37
Hens						4				6	
Plts.	13	36	42		36	4	8	16	16	31	37
Hens								2		6	
Plts.	14	48	37		27	3	3	18	20	32	38
Hens						6		1		6	
Plts.	15	40	31		28	5	11	14	15	26	32
Hens						1		3		7	
Plts.	16	41	37		31	4	5	20	23	24	31
Hens						1		2		7	
Plts.	17	51	38		22	4	5	19	21	29	36
Hens						4		3		8	
Plts.	18	41	38		28	6	10	21	24	26	34
Hens						2		2		8	
Plts.	19	48	39		18	5	7	23	25	28	36
Hens						2		4		5	
Plts.	20	47	33		21	7	9	24	28	32	37
Hens						3		5		13	
Plts.	21	47	44		22	6	9	22	27	21	34
Hens						2		2		4	
Plts.	22	43	38		17	8	10	28	30	31	35
Hens						2		10		5	
Plts.	23	45	33		15	6	8	22	32	29	34
Hens						1		3		9	
Plts.	24	49	33		14	8	9	25	28	26	35
Hens						2		8		8	
Plts.	25	47	34		13	6	8	34	42	18	26
Hens						3		5		7	
Plts.	26	40	29		10	6	9	34	39	22	29
Hens						1		9		6	
Plts.	27	48	39		11	8	9	30	39	18	24
Hens				13		3		9		6	
Plts.	28	44	31	1	14	7	10	30	39	23	29
Hens				8		3		9		6	
Plts.	29	45	38		8		8	30	39	13	19
Hens				8		3		8		8	
Plts.	30	46	32	1	9	7	10	36	44	15	23
Hens										15	
Plts.	31	44	34			7	7			10	25
Tot. Hens					717	107		108		247	
Tot. Plts.					2	130		560		810	
Grand Total		1,299	1,192		719		237		668		1,057

PART II

The Revolution and Its Results

CHAPTER V

Basic Theories

"Whys? and Wherefores!"

The writer, in his experience of twelve years of keeping poultry on the same piece of ground, in addition to his previous experience elsewhere, had reasons to ask a good many "whys," and has concluded that, as a result of asking the "whys" he now is in a position to write "wherefores" to some of these "whys."

Observing that he always got **some** winter eggs when many others, and among them some neighbors in close proximity, failed to get them, the question arose "why?" Their flocks looked well, but did not produce—the "wherefore" of this "why" was found in the fact that the other flocks were not fed sufficiently to sustain themselves, and at the same time produce eggs. Their outlay for feed was a dead expense; whereas, if they had increased the quantity of feed they would, in all probability, have had this expense returned to them, plus a small profit.

Observing, in caring for his flock late at night in the winter time, after the birds had been on the roost some hours, that the lantern light brought the birds off the roosts to attempt feeding; and that he had, on many occasions, to turn down the lantern light to a minimum, before he could get the birds back on the roost, the question "why" naturally arose, and to "Ask the birds, their judgment is good."

Observing that when placing feed in the feed box for cockerels (which having been separated from the pullets, were fed in a separate pen) late at night by lantern light, they immediately got down off the roost to eat out of the box, the question "why" came up, and the answering "wherefore" was that they must be hungry, but had to wait for a light before they could see to eat.

Observing, that when placing feed in the same manner for these cockerels on bright moonlight nights, when no light was carried, he often found the cockerels at ten p. m. or later eating out of the box in the moonlight—he decided that the "wherefore" to this "why" must be found in the fact that, while fowls could not see to find and pick up food in the dark, or by moonlight if scattered around in litter, they had sense enough to know that there was feed in the box which they could pick up without having to hunt, or look for it.

Limited Fuel Boxes

The author is endeavoring, in writing this book, to take the reader through something like the same process of reasoning whereby he arrived at his conclusions, and the causes which induced him to carry on further experiments.

If we wish to heat a large house or building, we invariably either study the question ourselves, or have experts study it for us, in order that the fuel boxes of our heating plant shall be adequate to meet all possible requirements, under the conditions we expect to meet, in the way of temperature variations. On top of these possible or probable requirements, we make provision with a surplus space in the fuel boxes, to care for any possible emergencies, defects, or leakages, in order that our heating plant may be at all times adequate.

Now let us take a hen's crop and make a little study of it in this light. Instance: a hen in a wild state, or in the semi-wild state, in which they are expected to produce eggs on the farms in the United States, we find that such a hen picks up an item here and an item there—in fact is continually at it, picking up, or searching for opportunities for picking up things. The bulk of the eggs produced in this country are produced on farms, according to statistical records, and farm conditions are mostly such as are now being described.

The hens on farms have their liberty, and in spring and early summer are able to pick up a good living, and produce a comparatively heavy crop of eggs at this season; their fuel boxes are adequate for the variations in temperatures and other drains on their vitality, which they must cope with at this season.

Later in the season, that is in late summer and early fall, the fuel boxes are still adequate; but the supply of fuel has fallen off, and the hens have to wear themselves out in their efforts to obtain enough to eat to sustain themselves and produce.

In the fall and winter the fuel supply is scarce indeed. The fowls are now mostly limited to what they can pick up indoors or around the barns; and generally farmers kindly throw them a few handfuls of corn, or other grain, at intervals more or less frequent, if they happen to think of it.

As a result of these conditions winter eggs on farms are, as a general thing, an impossibility.

The poultryman takes better care of his flocks, and for a great number of years has been endeavoring to supply all the fuel the fuel boxes could take care of—that is, he has been trying to similate spring and early summer conditions. He has, generally speaking, however, failed to realize that he was not keeping up the right amount of steam for efficient work—that he was not feeding his fuel boxes to their capacity.

The author's experience, however, was not of this nature. He came to the conclusion, some years ago, that a hen's fuel box must be taxed to the limit by crowding in fuel, and that of the best quality, to produce eggs in winter—and he acted on his conclusions with fairly satisfactory results.

Having done all he could to provide satisfactory environments for his flock, and having a flock which, by heredity, were good layers, and layers of winter eggs, he was still blocked in his efforts; so he came to the conclusion that the hen's fuel boxes were too small for their other possibilities of producing eggs, at all seasons of the year, under natural environments or conditions.

The Year Book of the Department of Agriculture for 1910, page 462, shows the importance and the need of "A Revolution in Egg Production." "Such climatic conditions as prevail during March and April in the Central States, both East and West, are ideal for egg production and egg marketing. Hence, it is only necessary to know the climate of a region in order to know when the egg supply is greatest and best. If one considers the number of months each year when climatic conditions preclude egg production almost entirely over the whole of our great egg producing territory, it is plain that some provision for these months of scarcity must be made from the season of plenty if eggs are to appear the year round on the tables of any except wealthy people. The development of the resources of Kentucky and Tennessee will help to ease the demand of the Eastern markets for "best fresh" eggs during the winter months, but it can never satisfy the general demand any more than the Northern belt, as represented by Michigan and Minnesota, can keep all supplied during the heat of midsummer. Therefore, we must continue to study, and work for, and urge, increased egg production wherever the little feathered lady can manage to eke out a living by dint of hard scratching, be it North, South, East or West. And we must remember, too, that every new laid egg is fresh, sweet, nutritious food. * * * * * The hen has kept pace with her breed and her environment, and almost invariably, even under the worst conditions, she has given her owner more than she receives."

Value of Light

We have stated, in our preface, that heredity and environment are two big factors in egg production.

Heredity is the sum of the effects of the environments of past generations, and this part of the subject is too comprehensive to take up in this little work.

We should, however, profit by the laws of heredity sufficiently to produce, or purchase, our foundation stock from sources which are known to have produced the kind of stock we desire.

We purchase farm stock, of other kinds, on this basis and theory that like begets like; and why should we not do the same with poultry?

A flock, produced from the eggs of stock of good laying quality, will almost surely produce heavier, under like conditions than flocks produced from eggs of indifferent layers.

Taking this position for granted, however, the second factor in production, namely, environment, will have more effect on laying qualities than heredity.

Flocks may be found which are excellent layers, and other flocks which are very poor layers, in cases where both kinds of flocks are produced from eggs obtained from the same source, and under identical conditions.

Nothing in the world but the effect of environment can account for such cases. It all depends on "the man behind the gun." One man hits the bull's eye—the other shoots wide of the mark, in providing environment.

In other words, we must supply such environments as will enable the stock to back up their reputation, and prove out on their heredity.

This little book, up to this point, has been dealing mostly with this second factor, of environment. The book, however, would never have been attempted, except as a result of a study of the third factor mentioned in the preface, which is "the length of the hen's business day."

The author had noticed for several years that, irrespective of extreme temperatures, the egg yield fell off in late fall and winter as the days got shorter, and that this falling off occurred even faster during cloudy and dark seasons, which were comparatively mild, than during bright sunshiny seasons of colder temperatures.

This condition was so obvious that the author decided that he would, at some time, experiment with an artificial day.

A Hen's Business Day

On June 21st, in the latitude of Chicago and Boston, the sun rises at about four twenty-three a. m. and sets at about seven forty p. m., making a day of fifteen hours and seventeen minutes; and in the same latitude, on December 21st, the sun rises about seven twenty-six a. m. and sets about four thirty-one p. m., making a day of nine hours and five minutes. This is a difference in the length of a hen's business day of six hours and twelve minutes.

This variation in the length of a hen's business day only tells a part of the story. The days in summer are nearly always light enough for the hens to transact business for their whole length; but many of the short days of winter will be so dark and dreary as to restrict the hens to a business day of seven hours or less.

This is a serious proposition, if we look at it from the viewpoint of the hens' capacity for factory production in the output of eggs. If we were running a factory, with men as the producers, we would at once equalize these conditions by artificial light, so as to make a uniform or nearly uniform day throughout the year.

With human factors, however, actual production ceases at the business hours, but, with the hens production goes on after business hours, but is not delivered, except in rare cases when they lay on the roost, until the following day.

In the short days of winter the fuel boxes get empty and the steam runs down. It takes more steam to keep going in December than in June; but, up to the present time, we have not taken means to provide this extra steam. The business day is restricted, the fuel boxes get empty, steam runs down, and we have to start new fires every morning; whereas, we should bank the fires ready for an early steam pressure the following day.

This is about how the author of this little work figured out the proposition, and decided that sometime, in the undetermined future he would try artificial light, to make a longer day, as a solution of the problem of satisfactory egg production.

The days in the winter 1913-1914 proved so dark and dreary that he decided to make an immediate start; and on January 21st, 1914, electric lights were installed in the poultry houses. We decided to "Ask the birds, their judgment is good."

CHAPTER VI

Remarkable Results

The Problem Solved.

The installation of the electric light in two poultry houses, which included the wiring, and the moving of the meter from the second story of the house to the basement, cost thirty dollars. Two sockets were placed in each poultry house, one socket in each house being used for a single lamp, and the other one being provided with a Benjamin socket.

In the Benjamin socket, in each house, an eight candle power incandescent lamp was placed on one side and a sixty candle power Tungsten lamp was installed on the other side, and in the other socket another sixty candle power Tungsten lamp.

The reason for providing Benjamin sockets was, that we wanted the small light to remain burning, to simulate dusk, while the fowls were going to roost, after turning down the larger lights.

Double wiring, so as to have the smaller lamps on a separate circuit and switch, would prove more convenient, as it would enable us to turn the larger lights out, and leave the smaller lights on, without having to go to the poultry houses.

For the purposes of this experiment, it was thought too expensive to put in double wiring, so we unscrewed the larger light in the Benjamin socket, and turned the other light out, leaving the smaller or eight candle power lamp burning until the fowls had all gone to roost. We then turned off the switch in the house, screwed up the loosened lamp in each Benjamin socket, and turned the other individual lamp switches on—thus leaving the lights already set for an early morning light, by turning the switch in the house at any time desired.

We found that the eight candle power lamp was too good a light to leave for the fowls to go to roost. We went visiting one evening, and left the smaller lights going, and when we came home at eleven p. m. found the birds as busy as at noonday.

We, therefore, substituted two candle power lamps for the smaller lights, and got over this difficulty.

At the time we turned on the electric light, we had received for that day and eleven days previous eleven, nineteen, eighteen, seventeen, eighteen, thirteen, seventeen, twenty-six, twenty-one, twenty-one, twenty-one, and twenty-six eggs respectively, a total of two hundred twenty-eight eggs in twelve days; and in the twelve days after turning on the light we received eggs as follows: Twenty-two, twenty-nine, twenty-nine, thirty-four, thirty-seven, forty-three, forty-six, fifty-four, forty-six, sixty-one, seventy-three, and on February 2nd eigthy-three, a total of five hundred fifty-seven for twelve days; which is over double the number of eggs for the twelve days immediately preceding.

The other conditions remained the same. The hens had plenty of feed and water, always available, before them, if the light was strong enough to see to get it. By merely lengthening the hen's business day we doubled the output in twelve days. We must also make allowance, for anyhow three or four days, while the fuel boxes were enabling the hens to replenish their bodily conditions.

We now use the same candle power in the larger house as when the light was first installed; but we use only one Tungsten lamp, of one hundred candle power, in the smaller house in one socket, and the small two candle power lamp in the other socket.

A reference to the laying record on Page 82 will show 1,943 eggs laid in twenty-eight days of February, an average of 69.39 eggs per day. There were 160 members of the flock for 13 days, $160 \times 13 = 2,080$ single hen days, and 159 members for the other 15 days, $159 \times 15 = 2,385$ single hen days, making 4,465 total single hen days; and this divided into the number of eggs—1,943—gives an actual average percentage of 43.52 per cent of an egg a day for each member of the flock.

This month of February was extremely cold with us—on five days the temperature reading touched from one to nine below zero Fahrenheit, and twelve days showed a maximum temperature during the day of twenty-five degrees, and less, Fahrenheit on a registering thermometer.

On ten days at seven a. m. the temperature registered less than five degrees above zero.

The reader can imagine that we felt elated at the result of our experiment, and the solving of the problem of winter egg production.

Given proper care, a balanced ration, liberal and regular supply of feed, and we had added only one factor of summer conditions, and that a factor heretofore not considered—we had merely lengthened the hens' business day to enable them to eat sufficient food to repair bodily waste, supply heat and energy, and leave a surplus for the production of eggs; and presto! we gathered one thousand nine hundred forty-three eggs in a February of extremely low temperature, as compared with one thousand ninety-five eggs for the previous February when the thermometer only touched zero twice, and that only for a short time. While we had some cold nights in the February of 1913, it warmed up in the day time as compared with February, 1914.

The birds continued to produce well, as will be seen by the production records on Page 83; and a comparison will prove interesting with the production record on Page 64 for 1913.

In due course arrived the moulting season. Few poultry-men expect many eggs at this season, and farm kept poultry seldom, if ever, produce at this season. Many poultrymen, and some poultry papers, maintain that hens cannot moult and lay at the same time.

It will be noticed in the production records on Page 84 that production steadily decreased; and, in discussing this problem with a friend, the author made the following statement of his position: "The birds had been doing well all summer, and were now moulting heavily, and were undergoing the greatest strain on their vitality; that the days were also shorter considerably than they had been—hence the hens had less opportunity to repair waste and energy and grow feathers; that he believed turning on the light, and lengthening the day at this time, would help the hens recuperate and get through the moult quicker. Also, that if they did this, they would probably produce during the coming winter as well as the pullets, under the improved conditions of the longer business day."

The light was turned on, on the 21st of August. The family at this time was away in Michigan, and did not return until September 5th. On this account the daily mash had been fed to the hens in early morning; but, after turning on the light, this procedure was changed, and it was fed at seven thirty p. m. after arriving home from business.

The five days previous to and including August 21st produced an output of eighty-nine eggs, and the five days thereafter produced seventy-seven eggs; and from there on the birds started in to produce again, and produced one thousand one hundred seventy-three eggs in September as against seven hundred nineteen for the September previous; and produced one thousand five hundred twenty eggs in October as against two hundred thirty-seven eggs in October, 1913—this in the face of the fact that in August, 1914, we received only eight hundred sixty-six eggs as against one thousand one hundred ninety-two eggs in August, 1913.

We had once more acted on the suggestion to "Ask the birds, their judgment is good."

We received eggs which will compare as follows:

Winter of 1913-1914			Winter of 1914-1915		
Month	Hens	Pullets	Month	Hens	Pullets
September	717	2	September	1,156	17
October	107	130	October	1,191	329
November	108	560	November	822	421
December	247	810	December	556	651
January	191	586	January	789	694
Totals	1,370	2,088	Totals	4,514	2,112

There were about the same number of hens, at this period, in each year; but there were less pullets in 1914 so that both the hens and the pullets profited in production by the longer days.

As stated before, the pullets always did fairly well in the winter season.

In January, 1914, when the lights were first installed, they were turned on from about six a. m. until seven or seven thirty a. m.—depending upon whether the morning was dark or bright, and were turned on at dusk until about seven thirty p. m. On February 28th, 1914, the whole family were in Chicago, so the lights were left going, and when we arrived home at eleven fifty-five p. m. the **egg factory** was still going full blast.

The winter of 1914-15 the lights had been turned on at about the same time in the morning; but had been left going until nine to nine fifteen p. m. before turning out the larger lights, in order to get the fowls on the roosts.

It is comical, when turning out the lights this late, to notice several of the birds get busy picking up grain by the two candle power light, after a sudden reduction from one hundred two and one hundred twenty-two candle power lights respectively in each poultry house.

Production Records for 1914-1915

Following are the production records for the year 1914 and the start of 1915. First, a table showing the items of feed purchased by months, and then the eggs produced for each day, and in monthly columns, with the pullet production separated from that of the hens; and then a table showing the total eggs produced each month and their value.

It will be seen that as a summary we have for 1914:

Weight of feed, 16,685 pounds; cost of feed, $333.24; number of eggs, 18,332; value of eggs, $560.06; meat sold, $49.56.

If we allow the meat sold to offset the feed used in raising the young stock, and the cost of keeping the cockerels until disposed of, together with cost of eggs for hatching, at market price for table eggs, we have a showing, as a rough deduction, of a cost for feed of 21.81c per dozen for 1,527 8-12 dozen eggs; and that it took 10.92 pounds of feed to produce a dozen eggs.

The same houses and yard were used as in the previous year.

The eggs were sold at an average price of 36.66c per dozen, this being 14.85c per dozen above the cost of feed.

The average cost of feed per hundred pounds in 1914 was $2.00.

The following summary will serve to make an easy comparison for the two years:

1913—Quantity of feed, 16,439; average cost per 100 pounds, $1,85; cost of feed, $303.34; number of eggs, 14,729.

1914—Quantity of feed, 16,685; average cost per 100 pounds, $2.00; cost of feed, $333.24; number of eggs, 18,332.

1913—Value of eggs, $403.96; number of dozens, 1,227 5-12; pounds of feed per dozen eggs, 13.39; cost of feed per dozen eggs, 24.71c.

1914—Value of eggs, $560.06; number of dozens, 1,527 8-12; pounds of feed per dozen eggs, 10,92; cost of feed per dozen eggs, 21.81c.

1913—Average price per dozen eggs, $32.91c; selling prive above feed cost, 8.20c; meat sold, $35.56.

1914—Average price per dozen eggs, 36.66c; selling price above feed cost, 14.85c; meat sold, $49.56.

Excess Figures 1914 Over 1913.

Pounds of feed, 246; cost of feed, $29.90; number of eggs, 3,603; value of eggs, $156.10.

It will be seen from the foregoing, that, for 246 pounds of feed we received 3,603 eggs, which, figured at the same price of 32.91c per dozen as in 1913, figures up to $98.81. This result is an offset to the cost of this 246 pounds of feed, at 1913 prices, of $4.55, plus the cost of electricity.

The cost of the extra light current consumed must, of course, be taken into account, The author has no means of arriving exactly at this cost, as the current for residence use and poultry houses went through the same meter. The total bills for current for the year 1913, however, were $34.75, and for the year 1914 $49.11, showing an excess of $14.36 for the year.

We used an electric iron for doing the domestic ironing in 1914, which consumed considerable current. As near as can be judged, therefore, the excess bill, for poultry houses, was about $12.00.

The light bills would have been considerably higher were it not for the fact that we came in on a lower rate, for excess current. These figures are for practically eleven months, as the electric light was not used until January 21st, 1914.

There were other items of expense which would only figure up to a very few dollars, outside of the expense for egg cartons which came to about $12.00. The comparison between the two years would not be affected by these items.

This experiment being a hobby, to which only spare time was devoted, no labor can be charged up—although this item would have to figure in a business plant. On this point the author is of the conviction that, with the labor saving conveniences and devices, suggested in another chapter, ten times as many birds could be well cared for with not more than one hour's time per day additional to that given to the care of this flock.

FEED PURCHASED IN THE YEAR 1914.

Date		Weight in Pounds	Variety	Total
January	10	500	Scratch Feed	$ 9.00
"		50	Beef Scrap	1.63
"		100	Scratch Feed	1.85
"	15	100	Cabbage	3.00
"	26	64	2 bus. Oats	.96
"		194	Mash	2.63
"	31	100	Scratch Feed	1.85
Total		1,108		$20.92
February	2	500	Scratch Feed	9.00
"		100	Mash	2.25
"		50	Beef Scraps	1.63
"	6	200	Yellow Turnips	3.00
"	9	50	Beef Scraps	1.63
"	16	500	Scratch Feed	9.00
"		100	Mash	2.25
"	17		Bale of Straw	.40
"		64	2 bus. Oats	.96
"	20	100	1 bbl. Cabbage	4.00
Total		1,664		$34.12
March	2	100	Scratch Feed	1.85
"		500	Scratch Feed	8.90
"		100	Mash	2.25
"		100	Oyster Shells	.80
"		100	Charcoal	2.25
"	5	50	Beef Scraps	1.63
"	9	300	Scratch Feed	5.34
"		100	Mash	2.25
"	12	100	Yellow Turnips	2.18
"	17	100	Mash	2.25
"	17	192	6 bus. Oats	2.88
"		100	Scratch Feed	1.85
"	25	500	Scratch Feed	9.00
"	25	25	Beef Scraps	.88
"	30	200	Mash	4.50
Total		2,567		$48.81

Date	Weight in Pounds	Variety	Total
April	2	50.........Beef Scraps....................	1.63
"	7	500.........Scratch Feed...................	9.00
"		100.........Mash	2.25
"	21	500.........Scratch Feed...................	9.00
"		50.........Beef Scraps....................	1.63
"	23	100.........Yellow Turnips.................	1.50
"	25	100.........Mash	2.25
Total ...1,400			$27.26
May	1	.........1 bale Straw.....................	$ .35
"		100.........Chick Feed...................	2.50
"		100.........Fine Grit.....................	.80
"		100.........Oyster Shells.................	.80
"	7	100.........Mash	2.25
"	19	100.........Mash	2.25
"		100.........Scratch Feed...................	1.85
"	20	128.........4 bus. Oats....................	2.30
"	25	100.........Scratch Feed...................	1.85
Total 828			$14.95
June	1	200.........Scratch Feed...................	3.70
"		50.........Beef Scraps....................	1.63
"		100.........Mash	2.25
"	10	100.........Chick Feed	2.25
"	13	100.........Scratch Feed...................	1.85
"	20	100.........Scratch Feed...................	1.85
"	25	100.........Scratch Feed...................	1.85
"	26	50.........Mash	1.13
Total ... 800			$16.51
July	1	50.........Beef Scraps....................	1.63
"		500.........Scratch Feed...................	9.00
"		100.........Mash	2.25
"	9	128.........4 bus. Oats....................	2.00
"	10	100.........Mash	2.25
"		100.........Grit.	.80
"	18	100.........Mash	2.25
"		50.........Beef Scraps....................	1.63
"	24	100.........Scratch Feed...................	1.85
"	28	100.........Scratch Feed...................	1.85
Total ...1,328			$25.51

Date		Weight in Pounds	Variety	Total
August	1	500	Scratch Feed	9.00
"		100	Mash	2.25
"		50	Beef Scraps	1.63
"	13	100	Mash	2.25
"	20	64	2 bus. Oats	.96
"		50	Mixed Meal	.85
"	22	100	Scratch Feed	2.39
"	25	100	Scratch Feed	2.39
"	27	100	Mash	2.25
Total		1,164		$23.97

Date		Weight in Pounds	Variety	Total
September	2	200	Scratch Feed	$ 4.20
"		50	Mixed Meal	.88
"	3	100	Oyster Shells	.80
"		50	Beef Scraps	1.63
"		100	Mash	2.50
"	9	200	Scratch Feed	4.20
"	16	500	Scratch Feed	10.50
"		50	Mixed Meal	.88
"	21	25	Beef Scraps	.83
"	25	100	Yellow Turnips	1.50
"	26	50	Mixed Meal	.88
"		64	2 bus. Oats	1.10
"	28	100	Mash	2.50
Total		1,589		$32.40

Date		Weight in Pounds	Variety	Total
October	1	50	Beef Scraps	1.63
"		100	Scratch Feed	2.10
"	5	50	Alfalfa Meal	2.00
"	10	100	Mixed Meal	1.94
"	12	200	Scratch Feed	4.30
"	14	100	Scratch Feed	2.10
"		50	Beef Scraps	1.63
"	17	100	Mash	2.50
"	22	200	Scratch Feed	4.20
"	24	128	4 bus. Oats	2.12
"		50	Mixed Meal	.95
"	28	100	Scratch Feed	2.10
"	29	25	Beef Scraps	.88
"		50	Mash	1.25
"		10	Beef Scraps	.40
Total		1,313		$30.10

Date	Weight in Pounds	Variety	Total
November 2	100	Mash	$ 2.50
"	50	Beef Scraps	1.63
"	100	Grit	.78
"	100	Crate of Cabbage	1.50
"	200	Scratch Feed	4.10
" 9	100	Scratch Feed	2.10
" 11	100	Scratch Feed	2.10
"	100	Mash	2.50
" 16	50	Mash	1.25
" 20	60	1 bbl. Bread	.75
"	50	Beef Scraps	1·63
" 23	200	Scratch Feed	4.20
"	100	1 bbl. Cabbage	1.35
" 25	50	Mash	1.25
"	4	Plucks	.12
Total	1,364		$27.76
December 1	500	Scratch Feed	$10.00
"	50	Mash	1.25
" 7	50	Beef Scraps	1.63
" 8	100	Mash	2.50
"	100	Oyster Shells	.80
" 9	100	1 bbl· Cabbage	1.75
" 12	60	1 bbl. Bread	.75
" 14	500	Scratch Feed	10.00
" 17	100	Mash	2.25
Total	1,560		$30.93

EGG RECORD

	Day	1914—January	Total	February	Total	March	Total		April Total	May Total	June Total
Hens		5		24		20		44			
Plts.	1	14	19	49	73	43	63	38	82	67	69
Hens		6		40		21		35			
Plts.	2	3	9	43	83	35	56	34	69	62	58
Hens		5		30		26		66			
Plts.	3	13	18	41	71	46	72	42	108	56	58
Hens		3		31		24		44			
Plts.	4	13	16	53	84	37	61	38	82	54	55
Hens		6		38		29		38			
Plts.	5	12	18	47	85	37	66	39	77	54	58
Hens		3		34		22		46			
Plts.	6	13	16	36	70	40	62	28	74	62	60
Hens		9		50		19		56			
Plts.	7	11	20	36	86	40	59	40	96	53	57
Hens		2		27		33		30			
Plts.	8	13	15	44	71	49	82	37	67	54	56
Hens		4		28		23		33			
Plts.	9	13	17	42	70	42	65	37	70	53	60
Hens		1		33		26		44			
Plts·	10	10	11	42	75	37	63	41	85	79	63

	Day	1914—January	Total	February	Total	March	Total		April Total	May Total	June Total
Hens		7		26		20		34			
Plts.	11	12	19	50	76	41	61	24	58	53	54
Hens		5		27		20		55			
Plts.	12	13	18	37	64	39	59	48	103	58	58
Hens		4		25		22		48			
Plts.	13	13	17	41	66	42	64	37	85	55	48
Hens		4		28		22		40			
Plts.	14	14	18	42	70	39	61	32	72	62	74
Hens		3		22		28		36			
Plts.	15	10	13	42	64	40	68	41	77	59	57
Hens		4		21		23		48			
Plts.	16	13	17	43	64	39	62	39	87	53	56
Hens		7		27		34		42			
Plts.	17	19	26	43	70	41	75	37	79	68	55
Hens		2		43		33		45			
Flts.	18	19	21	18	61	39	72	39	84	56	52
Hens		5		41		41		46			
Plts.	19	16	21	32	73	41	82	40	86	52	44
Hens		2		20		42		44			
Plts.	20	19	21	44	64	38	80	37	81	61	48
Hens		6		21		41		40			
Plts	*21	20	26	48	69	41	82	40	80	54	51
Hens		2		25		50		28			
Plts.	22	20	22	47	72	40	90	44	72	55	46
Hens		7		22		53		42			
Plts·	23	22	29	39	61	31	84	40	82	74	51
Hens		7		20		43		41			
Plts.	24	22	29	40	60	48	91	40	81	66	58
Hens		10		28		51		28			
Plts.	25	24	34	40	68	37	88	35	63	61	55
Hens		9		21		43		43			
Plts.	26	28	37	41	62	49	92	46	89	69	49
Hens		9		19		46		43			
Plts.	27	34	43	40	59	41	87	42	85	66	56
Hens		10		14		38		——			
Plts.	28	36	46	38	52	39	77	1.139	68	65	59
Hens		16				42		1,035			
Plts.	29	38	54			46	88		67	60	44
Hens		8				44					
Plts.	30	38	46			43	87		60	66	41
Hens		20				41					
Plts.	31	41	61			38	79			51	
Tot. Hens		191		785		1,020					
Tot. Plts.		586		1,158		1,258					
Grand Total			777		1,943		2,278		2,369	1,858	1,650

*Electric light turned on this date.

	Day	1914—July Total	Aug. Total	September Total	October Total	November Total	December Total
Hens					38	32	19
Plts.	1	57	37	28	6 44	15 47	19 38
Hens					40	34	20
Plts	2	44	39	28	1 41	14 48	20 40
Hens					40	31	15
Plts·	3	53	44	34	5 45	14 45	29 44
Hens					35	34	22
Plts.	4	36	30	39	5 40	16 50	22 44
Hens					42	37	26
Plts.	5	54	40	36	7 49	14 51	19 45
Hens					48	28	29
Plts.	6	36	40	45	7 55	12 40	17 46
Hens					42	40	15
Plts.	7	49	34	32	5 47	18 58	23 38
Hens					44	38	16
Plts.	8	50	36	41	6 50	12 50	30 46
Hens					34	33	20
Plts.	9	40	41	41	11 45	15 48	25 45
Hens					48	21	15
Plts.	10	52	33	37	4 52	19 40	24 39
Hens					38	37	19
Plts.	11	45	45	28	11 49	15 52	26 45
Hens					48	30	19
Plts.	12	52	29	53	10 58	12 42	26 45
Hens					37	31	18
Plts.	13	52	45	40	11 48	18 49	22 40
Hens					38	36	15
Plts.	14	47	27	33	5 43	12 48	29 44
Hens					41	35	14
Plts.	15	48	33	30	16 57	12 47	20 34
Hens					44	32	14
Plts.	16	61	29	42	10 54	12 44	20 34
Hens					34	37	17
Plts.	17	50	24	40	11 45	11 48	22 39
Hens					35	26	17
Plts.	18	53	23	43	10 45	14 40	20 37
Hens					39	31	16
Plts.	19	54	17	42	12 51	16 47	17 33
Hens					29	23	16
Plts.	20	45	14	36	15 44	10 33	20 36
Hens					38	18	21
Plts·	21	56	11*	42	13 51	14 32	13 34
Hens					41	18	18
Plts.	22	57	13	36	13 54	12 30	17 35
Hens				44	28	25	12
Plts.	23	39	17	1 45	17 45	11 36	21 33
Hens				42	44	17	19
Plts.	24	39	12	42	16 60	10 27	21 40

	Day	1914—July Total	Aug. Total	September Total		October Total		November Total		December Total	
Hens				42		36		22		15	
Plts.	25	39	19	3	45	17	53	16	38	14	29
Hens				40		39		13		16	
Plts.	26	29	16	2	42	10	49	13	26	21	37
Hens				44		30		17		15	
Plts.	27	36	18	2	46	19	49	16	33	18	33
Hens				40		35		16		19	
Plts.	28	55	24	3	43	17	52	15	31	19	38
Hens				40		30		17		17	
Plts.	29	36	23	3	43	13	43	11	28	20	37
Hens				38		42		13		21	
Plts.	30	50	26	3	41	11	53	22	35	18	39
Hens				34						21	
Plts.	31	34	27			15	49			19	40
Tot. Hens			1,156		1,191		822		556		
Tot. Plts.			17		329		421		651		
Gd. Tot.		1,448	866		1,173		1,520		1,243		1,207

*Electric light turned on this date.

EGG RECORD
January, 1915

	Day	Hens	Pullets	Total
Hens / Pullets	1	19	20	39
Hens / Pullets	2	21	20	41
Hens / Pullets	3	25	21	46
Hens / Pullets	4	20	19	39
Hens / Pullets	5	18	21	39
Hens / Pullets	6	25	21	46
Hens / Pullets	7	27	25	52
Hens / Pullets	8	17	18	35
Hens / Pullets	9	21	30	51
Hens / Pullets	10	35	22	57
Hens / Pullets	11	21	27	48
Hens / Pullets	12	19	25	44

	Day	Total
Hens		35	
Pullets 13	25	60
Hens		26	
Pullets 14	24	50
Hens		34	
Pullets 15	30	64
Hens		30	
Pullets 16	22	52
Hens		35	
Pullets 17	27	62
Hens		39	
Pullets 18	19	58
Hens		33	
Pullets 19	24	57
Hens		32	
Pullets 20	22	54
Hens		29	
Pullets 21	21	50
Hens		25	
Pullets 22	20	45
Hens		22	
Pullets 23	21	43
Hens		28	
Pullets 24	22	50
Hens		22	
Pullets 25	17	39
Hens		27	
Pullets 26	24	51
Hens		16	
Pullets 27	22	38
Hens		26	
Pullets 28	19	45
Hens		25	
Pullets 29	22	47
Hens		12	
Pullets 30	21	33
Hens		25	
Pullets 31	23	48

| Total Hens................ | 789 |
| Total Pullets.............. | 694 |

| Grand Total .... | 1,483 |

RECAPITULATION FOR 1914

Month	Weight of Feed Purchased	Cost of Feed	Number of Eggs	Value of Eggs
January	1,108	$ 20.92	777	$ 31.46
February	1,664	34·12	1,943	70.27
March	2,567	48.81	2,278	83.06
April	1,400	27.26	2,369	57.96
May	828	14.95	1,858	45.10
June	800	16.51	1,650	41.35
July	1,328	25.51	1,448	36.18
August	1,164	23.97	866	23·09
September	1,589	32.40	1,173	38.57
October	1,313	30.10	1,520	45.25
November	1,364	27.76	1,243	39.24
December	1,560	30.93	1,207	48.53
	16,685	$333.24	18,332	$560.06
January (1915)	1,781	$35.69	1,483	$61.80

Six hundred ninety-five eggs were sold for hatching for $37.80.

MEAT SOLD IN 1914

Month		Items	Weight	Price	Value
April	20	1 Pullets	3 lbs.	$ .18	$ .54
"	"	10 Hens	33 "	.18	5.94
"	22	12 "	47 "	.18	8.46
"	27	12 "	48 "	.18	8·64
May	6	10 "	39 "	.18	7.02
June	20	4 Roosters	16½ "	.12	1.98
"	27	2 Hens	8 "	.15	1.20
July	3	1 "	4 "	.15	.60
"		1 Cockerel	1½ "	.18	.27
"	20	1 Hen	4 "	·15	.60
"	25	3 Cockerels	4½ "	.22	.99
August	24	8 "	12 "	.19	2.28
"		4 "	6 "	.19	1·14
"	29	2 "	3 "	.19	.57
September	12	3 "	6 "	.19	1.14
"	19	4 "	8 "	.18	1.44
	26	4 "	7½ "	.18	1.35
October	3	4 "	8 "	·18	1.44
"	10	3 "	7½ "	.18	1.35
"	17	3 "	8 "	.17	1.36
November	1	1 Rooster	4 "	Sold alive	1.25
Total			278½		$49.56

Longer Day Effects on Health and Vigor

The effects of a longer business day, on the health and vigor of the flock, are almost immediately noticeable. Within two or three days, the fowls seem to begin to fill out, their feathers begin to take on a gloss, and their combs begin to redden.

After a lapse of one week's time, under the new conditions, one would scarcely recognize them as being the same flock.

Enjoying the Lengthened Day

Combs gradually increase in size, and get a deeper red color. In a few days, the appearance of the flock shows the identical changes which occur, under natural conditions, when spring arrives, and the birds are able to get outdoors on the ground and on the grass.

Many, when first nibbling on this idea of a longer business day for the hens, seem to get the impression that the hens are **forced,** by the new conditions, into laying eggs, and that, therefore, they will soon **wear out.**

If we were to really attempt to force laying, by the use of drugs or condiments, such reasoning would likely prove correct. By lengthening the day, however, we are doing nothing of the kind—we are merely placing before the hens an opportunity to help themselves, at will, to whatever they need, for a longer period in each twenty-four hours.

As absolute proof that a longer day is conducive to health and vigor, rather than the reverse, we have here the records of a flock, under these so-called forcing conditions, yielding well in eggs in the winter of 1913-1914; and, with the longer day, an increasing yield from January 21st, until the natural longer days of spring, and then right on through the summer.

We now come to the time of all others when the lack of vigor, as a result of the supposed debilitating influences of the conditions under consideration would appear, namely, the moulting season.

A glance at the tables furnished, shows a **gradual** decrease in egg yield from August 11th until August 21st, when the birds were in such a heavy moult that the dropping boards were literally covered with feathers.

Other seasons, we would expect the yield to drop off entirely when the fowls were in the condition shown at this time. The combs would shrink, and lose nearly all their color, the fowls would stand around in a listless manner, as if they were tired of life, and altogether present a more or less bedraggled appearance.

Now was thought a good time to "Ask the birds, their judgment is good," as to whether a longer day, at this time, would help them obtain sufficient nourishment to sustain the drain on their systems, incident to growing a new crop of feathers, and at the same time lay eggs. To ask them whether, in their judgment, they were played out as a result of having had such long days in the past winter of **cruel** treatment, in being forced off their perches so early in the morning, and kept off their perches, by the glare of the light, so late at night.

Judging by results in winter and spring, we had a somewhat sure feeling of what the answer would be; but one never can be positive of such things, until they are put to **the actual** test of practical experience.

After the light had been turned on, the birds did not hesitate, by their actions and appearance, to herald their joy at the seeming return of spring. They took on new life, filled out in flesh, combs turned red and increased in size, and the

The Pullets in a Busy Night Scene

new feathers came in as if by magic; and, to show as proof that the long business days of the past winter and early spring had no evil effects on the hens, they commenced to increase in their egg yield, in the middle of a heavy moult, and not only laid right through the moult, as will be seen by the table, but the hens also laid as well all through the winter of 1914-1915, as did the pullets raised in 1914.

The author feels that a study of these records will prove, more conclusively than anything he could say, that a longer business day, not only does not tend to debility and deterioration, but also that, for a poultryman who is not afraid to feed his flock a sufficiency of balanced rations, a longer day is the one thing lacking and needed to preserve vigor, to prevent debility under trying conditions whether due to moulting or extremes of cold temperatures, and to enable his flock to produce eggs abundantly at all seasons.

The theory has been advanced, and used, that the moult can be forced, or brought on, by reducing the feed to a minimum for a short period; and afterward the period of moulting be shortened by increasing the quantity of feed and adding an extra supply of oily feeds such as sunflower seeds or linseed meal to help the fowls in growing their new coat of feathers.

The same reasons are responsible for the moult brought on, out of season, and against the poultryman's wishes; when early hatched pullets and hens moult at the beginning of winter.

The sudden call for increased feed, on the arrival of cold weather, catches the birds unprepared and unable to procure what they need—hence the moult.

With a longer business day, and plenty to eat, even the old hens do not seem to suffer from, or mind, the moult even in the dead of winter. We had about four such hens which had not moulted in the fall of 1914. These hens went about their business in zero weather losing feathers and making new feathers as if nothing unusual was transpiring.

Late hatched pullets will be in a position to mature, and develop winter layers, with the aid of the benefits derived from a longer business day.

End View of Pullet House

CHAPTER VII

Practical Application of Proven Theories

Production Possibilities

What are the possibilities in egg production? This is a very interesting question, and one which is hard to answer.

The records given herein for 1914 show an average production of 138.33 eggs from each member of the flock, based on an actual number of single hen days. These records show **this** as a possibility without question—that whole flocks can average 138.33 eggs a year. This average also counts in all the pullets as being mature at five months from hatching.

The conditions for producing in this flock are far from being the best or ideal. First, the limited quarters on a city lot preclude the possibility of raising enough young chicks, each year, to enable the owner to cull out the pullets as closely as would be advisable.

Few flocks of chickens develop evenly, under identical conditions, for all the individuals in the flock. Some seem to thrive out of all proportion to the others; others thrive fairly well, and still others only half as well as the condition of the best members of the flock would lead one to expect. Some few in the flock may seem to be continually far behind, in the race for development.

After the pullets have developed, and get near to laying age, and get their complete plumage, these differences are not so noticeable, except for the few really inferior birds.

First. If enough chickens could be raised to cull so closely as to leave in the flock only the birds showing exceptional vigor, at say three months from hatching, one could reasonably expect greater health and vigor in the flock, and, as a consequence, far greater returns in eggs.

Second. The condition of soil in the yards, which soil is a heavy clay loam, makes this soil pack very easily, interfering to a great extent with the opportunities for the birds to wallow, and exercise outdoors in the open season; also, in order to keep these yards in a sanitary condition, they must be frequently limed.

Third. The owner of the flock having to leave home before seven a. m., and not being able to arrive home again before seven thirty p. m., and often not before ten to eleven thirty p. m. week days (including Saturdays) his opportunities for putting into practice the principle of "Asking the birds, their judgment is good," were, of course, greatly limited.

At night the automatic feeders had to be filled for the next day, the dry mash, grit, oyster shell, and charcoal hoppers had to be seen to, and replenished when necessary, and the water fountains filled and put in place.

In the morning the green feed of sprouted oats, cabbage, or whatever was available was taken down, and the houses opened up before leaving home for the day's business.

In this way the work for the good wife was cut down to feeding the moistened mash, once a day, and gathering the eggs.

Business is business, and the keeping of poultry in the rear of the lot was only a diversion in an extended experiment, over a number of years, in the possibilities of a problem in economics. Business came first, and could never be neglected for a hobby.

We have read of individual hens producing two hundred eggs and over; and these records, many of them, come from reliable and authentic sources. We should always set our ideals high, in order to strive to attain them; and, if we succeed in actually reaching our ideals, we should not be satisfied, but should advance our ideals to a higher point—thus always having something ahead of us worth striving to achieve.

Much depends on the breed, and the different treatment needed to successfully handle each breed. With the heavier breeds, susceptibility to broodiness must be reckoned with as a hindrance and a handicap. With the Mediterranean breeds this handicap of broodiness is greatly cut down.

For laying purposes, the concensus of opinion in the United States seems to lean toward White Leghorns. These birds, under proper care and feeding, are certainly efficient "egg machines."

Judging by past experience, and taking account of the handicaps under which he has operated, the author has no hesitancy in placing the productive possibility of White Leghorns, when handled **as a business proposition,** as an average of two hundred eggs a year from each hen, in large flocks, under proper care, with liberal feeding of balanced rations, and with business days equalized in length, by artificial means, sufficiently to enable the hens to manufacture their product.

The figures given herein, on the comparative production for the years 1913 and 1914, were chosen, for this purpose, because the flocks matched better for these two years as to the number of birds and the corresponding ages of the birds making up the flocks. The two years being consecutive are also better for comparison.

As noted before, on Page 55, the birds in 1913 contained pullets, one and a half year old, and two and a half year old hens at the beginning of 1913; and pullets, one and a half year old, two and a half year old, and a few three and a half year old hens at the beginning of 1914. Thus there were a few hens in the flock over four and a half years old at the beginning of 1915.

The author has records of much better production in 1911, which follow, and these are given to show that for great production, under the best conditions with a natural day, **the age of the fowls has a good deal to do with the number of eggs produced.**

The tables for 1911 show an average production, for the whole flock, of one hundred fifty eggs. This flock contained pullets of 1910, and fifty-nine pullets hatched May 18th, 1911, and in this average, of one hundred fifty eggs, these 1911 pullets are figured in as being matured birds at four months and eighteen days old on October 1st, 1911.

The tables for 1911 show an average production for the adult fowls hatched May 30th, 1910, of one hundred seventy-eight eggs. All these fowls were nineteen months old at the end of the year of this record.

The reader will thus be in a position to note that even with the handicap of age of the adult fowls, and the lower average for the year 1914 as compared with that of the year 1911, **because of this handicap,** the fall and winter production of 1914-1915, with the aid of an artificial day, compares very favorably with the fall and winter of 1911-1912, and conclusively proves the value and aid of a longer business day.

The following table will show this quite clearly:

Adult Fowls

Fall and winter of 1911-12.
No hens in this flock over 20 months old at the end of this period.

	No. of Hen Days	No. of Eggs	Per Cent
Sept.	1,434	545	38 %
Oct.	1,457	367	25.19%
Nov.	1,410	201	14.25%
Dec.	1,445	273	18.89%
Jan.	1,426	93	6.52%
Totals	7,172	1,479	20.62%

Adult Fowls

Fall and winter of 1914-15.
Some hens in this record over 4½ years old at the end of this period.

	No. of Hen Days	No. of Eggs	Per Cent
Sept.	2,969	1,156	38.94%
Oct.	3,004	1,191	39.65%
Nov.	2,822	822	29.12%
Dec.	2,826	556	19.67%
Jan.	2,790	789	28.24%
Totals	14,411	4,514	31.32%

When it is remembered that the reasons, previously given, to show the advantages of a longer business day, demonstrate that these reasons merely add one (and that a heretofore neglected and essential) factor to the summer conditions, necessary to good egg production, the above comparison will serve to emphasize the great importance of this factor in profitable poultry keeping.

Not over thirty-seven per cent of the adult fowls in the fall and winter of 1914-1915 were of the same age as the adult fowls in the same period of 1911-1912—the other sixty-three per cent had a handicap of one, two, and some of them three years, to overcome.

It is reasonable to suppose that, if fowls of such ages were able to outstrip their youthful competitors during the moulting season and in the winter months, because of the advantages of a longer business day, the young stock would have made a much better showing than they did, for this same period, if the length of their working day had been equalized.

Taking these comparisons under consideration, the two hundred egg goal, previously mentioned as being possible, does not seem at all unattainable for whole flocks of young and vigorous birds under "A Revolution in Egg Production."

From an average of one hundred seventy-eight eggs per annum to an average of two hundred eggs per annum does not seem such a great increase, in the light of the results shown in the preceding chapters.

AVERAGES AND PERCENTAGES.

1911	No. of Hens		No. of Days		Single Hen Days	Total Monthly Hen Days	No. of Eggs	Per Cent	Remarks
January ..51	51	×	31	=	1581	1581	664	42 %	
February ..51	51	×	28	=	1428	1428	835	58.48%	
March 51	51	×	29	=	1479				1 Hen
	50	×	2	=	100	1579	1105	70 %	Out 3/2
April 50	50	×	30	=	1500	1500	1105	73.67%	
May 50	50	×	16	=	800				1 Hen
	49	×	15	=	735	1535	1021	66.51%	Out 5/6
June 49	49	×	30	=	1470	1470	837	56.94%	
July 49	49	×	31	=	1519	1519	915	60.23%	
August. ...49	49	×	22	=	1078				1 Hen
	48	×	9	=	432	1510	843	55.83%	Out 8/22
September .48	48	×	24	=	1152				1 Hen
	47	×	6	=	282	1434	545	38 %	Out 9/24
October									
Hens ...47	47	×	31	=	1457	1457	367	25.19%	
Pullets ..59	59	×	31	=	1829	1829	28	1.53%	
November									
Hens ...47	47	×	30	=	1410	1410	201	14.25%	
Pullets ..59	59	×	30	=	1770	1770	271	15.32%	
December									1 Hen
Hens ...47	47	×	19	=	893				Out 12/19
	46	×	12	=	552	1445	273	18.89%	
Pullets ..59	59	×	31	=	1829	1829	588	32.15%	
Totals						23296	9598	41.15%	
Totals for Adult Fowls only						17868	8711	48.75%	

NOTE—41.15% of an Egg a day for 365 days equals an average of 150 Eggs from all Birds in the flock.

48.75% of an Egg a day for 365 days equals an average of 178 Eggs from all Adult Fowls.

The Year Book of Agriculture for 1910 recognizes the importance and immensity of the value of our egg production in the aggregate. In a chapter on "The effect of the present method of handling eggs on the industry and the product," the following excerpts will give the reader an idea of the economic importance of "A Revolution in Egg Production," if such a revolution is generally carried on:

"During the calendar year 1909, 4,256,320 cases of eggs were received in the City of New York. Each case contained 30 dozen, hence there were 1,532,275,200 individual eggs, or enough to permit of a per capita consumption per annum of 321. * * * * According to the report of the Secretary of Agriculture for 1907, "More than $600,000,000 must be regarded as the value of the poultry and eggs produced on the farms in 1907. The amount may easily be larger. This industry has advanced at such a rapid rate that no arithmetic can keep up with it." Again in 1908, he says, "The eggs and poultry produced on the farms are worth as much as the * * * * * hay crop or the wheat crop," the latter being estimated at $620,000,000 for 1908.

"In eggs and poultry, then, we have an agricultural product of enormous money value, considered either individually or by comparison with other agricultural productions. * * * * * The output of eggs is steadily growing, but the demand is growing even faster than the supply, due to the increased price of meat, as well as a preference for eggs as food; hence, the price of eggs has gone up. In 1899 the farm price was 11.15 cents per dozen, as an average for the United States; in 1909 the average was 19.7 cents. * * * * * These are the prices to the producer, not the customer. The latter pays from fifty to one hundred per cent more than the producer receives. Some of the reasons for the increase to the consumer will be discussed in this article."

Averages and Percentages

There is an old saying that "figures don't lie;" but many so-called average and percentage records are wide of the mark, because they are based on erroneous methods of arriving at results. This is especially true as to the averages and percentages in egg production.

When a hen drops out of the race, she does not do so conveniently at the first of the month, but may ask the poultryman to dispense with her services, or she may quit by request, at any time during the month.

To arrive at the average number of hens in any one month, we must understand that we cannot either deduct a hen dropping out, say, on the 5th of the month, from our number of hens; nor can we leave such a hen to be accounted for in the figuring as having equal value, in the results, with those in the flock for the full month.

The only way we can get an exact percentage is to figure on the number of single hen days, and divide this number into the egg record, for the period for which we want the percentage.

This method is shown on Page 97.

Suppose we have a flock of one hundred sixty hens for the first ten days in a month, then sell off sixty hens, leaving one hundred hens for three days, and then sell off forty hens, we would have left sixty hens in the flock for the rest of the month.

The erroneous way sometimes used to figure this percentage, would be to add the numbers up for each period, and then divide by three (the number of periods) to get the average number of hens in the flock; thus $160+100+60=320$, and $320 \div 3 = 107$.

This figure would be multiplied by the number of days in the month, and then the result would be divided into the number of eggs to show the average percentage of eggs laid, by each hen, of a possible egg per day. Suppose we try this on a month of thirty-one days with an egg record of 1,192 eggs. We get a percentage of 35.94 per cent.

If we figure this correctly, we should multiply $160 \times 10 = 1,600$; $100 \times 3 = 300$; $60 \times 18 = 1,080$; and $1,600+300+1,080=2,980$ single hen days; which, divided into the number of eggs for percentage, in this case 1,192, would make an exact percentage of forty per cent. This shows a difference of over four per cent due to wrong method of figuring. The errors of this method may show errors as either more or less than the correct results.

Using this method for our 1914 production, we had results as follows:

AVERAGES AND PERCENTAGES FOR ADULT FOWLS IN 1914

1914	No. of Hens	No. of Days	Single Hen Days	Total Monthly Hen Days	No. of Eggs	Per Cent	Remarks
January 160	×	31 = 4960	4960	777	15.67%		
February ...160	×	13 = 2080					1 pullet out 2/13
159	×	15 = 2385	4465	1943	43.52%		
March 159	×	6 = 954					1 pullet out 3/ 6
158	×	4 = 632					1 hen out 3/10
157	×	8 = 1256					1 hen out 3/18
156	×	2 = 312					1 hen out 3/20
155	×	11 = 1705	4859	2278	46.88%		
April 155	×	15 = 2325					1 hen out 4/15
154	×	5 = 770					1 pullet out 4/20
143	×	2 = 286					1 hen out 4/20
131	×	5 = 655					12 hens out 4/22
119	×	3 = 357	4393	2369	53.93%		12 hens out 4/27
May 119	×	6 = 714					10 hens out 5/ 6
109	×	25 = 2725	3439	1858	54.03%		
June 109	×	27 = 2943					2 hens out 6/27
107	×	3 = 321	3264	1650	50.55%		
July 107	×	3 = 321					1 hen out 7/ 3
104	×	17 = 1768					1 hen out 7/17
103	×	11 = 1133	3222	1448	44.94%		
August 103	×	31 = 3193	3193	866	27.12%		
*Sub Totals			31795	13189			
September ..103	×	3 = 309					3 hens out 9/ 3
100	×	7 = 700					2 hens out 9/10
98	×	20 = 1960	2969	1156	38.94%		
October 98	×	18 = 1764					2 hens out 10/18
96	×	5 = 480					1 hen out 10/23
95	×	8 = 760	3004	1191	39.65%		
November .. 95	×	16 = 1520					1 hen out 11/16
94	×	7 = 658					2 hens out 11/23
92	×	7 = 644	2822	822	29.13%		
December .. 92	×	18 = 1656					2 hens out 12/18
90	×	13 = 1170	2826	556	19.67%		
Totals			43416	16914	38.95%		

NOTE—38.95% of an Egg a day for 365 days equals an average of 142.16 Eggs from each Adult Fowl in the flock, for the whole year 1914.

*These sub-totals are carried over to the next table.

SUPPLEMENTARY TABLE TO INCLUDE AVERAGES AND PERCENTAGES FOR PULLETS WITH THE ADULT FOWLS.

1914	No. of Birds	No. of Days	Single Hen Days	Total Monthly Hen Days	No. of Eggs	Per Cent	Remarks
September	.103 ×	3 =	309				3 hens out 9/ 3
	100 ×	7 =	700				2 hens out 9/10
	98 ×	20 =	1960				
	†		34	3003	1173	39.06%	
October	....‡152 ×	18 =	2736				
							2 hens out 10/18
	150 ×	5 =	750				1 hen out 10/23
	149 ×	8 =	1192	4678	1520	32.49%	
November	..149 ×	16 =	2384				1 hen out 11/16
	148 ×	7 =	1036				2 hens out 11/23
	146 ×	7 =	1022	4442	1243	27.98%	
December	..146 ×	2 =	292				1 pullet out 12/ 2
	145 ×	8 =	1160				1 pullet out 12/10
	144 ×	8 =	1152				2 hens out 12/18
	142 ×	9 =	1278				1 pullet out 12/27
	141 ×	4 =	564	4446	1207	27.18%	
Totals				16569	5143		
*Sub-totals to August 31, 1914				31795	13189		
Grand Totals				48364	18332	37.90%	

NOTE—37.90% of an Egg a day for 365 days equals an average of 138.33 Eggs from each hen or pullet in the flock, for the whole year 1914.

*These sub-totals are brought over from the previous table, to add in with this table, to show the averages and percentages for the whole flock.

†34 days added to cover 17 Eggs for Pullets beginning to lay.

‡54 pullets raised.

Press Comments on the Experiment.

When the author found that his experiment with a longer business day, for the hens, correlative and supplementary to his liberal and regular feeding of balanced rations, was a success, he decided that an economic fact, of such importance, should be given to the public, in tangible form, as soon as sufficient data had been accumulated to present a preponderance of results to make its success apparent, without opportunities for skepticism.

Having succeeded, where others had failed, in discovering the **real** reason for low egg production in fall and winter, by well cared for flocks, the author felt a pardonable pride in wishing the public to know the facts, and to be benefitted by his discovery.

Of course, an experiment of this kind could not escape a certain amount of local publicity, because such a new event as an "egg factory" running full blast after dark, could not very well be hidden from the view of passers-by on the street —either afoot, or in automobiles or other conveyances.

Through some channel, of which the author has no knowledge, the "Chicago Tribune" editorial staff was informed of the fact that the author was using electric light in his poultry houses; and they called him up on the phone, at his place of business, on January 4th, 1915, with a request for an interview on the subject.

Realizing that having started out to get information they would succeed in getting what they wanted in some manner, the interview was granted, and they sent a photographer out to take a few pictures.

On January 5th, 1915, the following pictures and write-up appeared in the "Chicago Tribune."

The pictures were taken by flash light, after dark, and show the "egg factory" in actual operation.

FOOLS CHICKENS; GETS MORE EGGS

G. G. Newell Installs Electric Light in Coops and Hens Work Overtime.

George G. Newell is an auditor. Figures and statistics and chickens are his hobbies. Efficiency is his watchword.

Back of his residence in Congress Park there is an inclosure forty feet square in which he keeps what he calls his "150 egg machines." The "machines" belong to the feathered tribe known as White Leghorns. He expects and obtains eggs from these "machines" with the same regularity and accuracy as he does figures from an adding machine.

Gets 18,000 Eggs.

He says he has obtained 18,000 eggs from his "machines" in the last year, or an average of an egg every third day for each fowl, and expects to bring this average up to an egg every other day for each hen during 1915. All the hens are laying now and he sells the eggs for 50 cents a dozen. Mr. Newell attributes his success to the fact that his chickens live in two electric lighted coops, go to roost by electricity, and get up at the beck of 100 candle power.

"I figured the whole problem out in black and white," said Mr. Newell. "I found that my chickens were not laying much in winter. They'd go to roost earlier in the winter months and get up later. I figured they didn't have sufficient daylight in which to eat the necessary amount of

Plays Electric Light Joke on Chicks and They Lay for It.

food and to get the required amount of exercise for good laying. I estimated they got about sixteen hours of daylight in midsummer and only about seven hours in midwinter. I decided to strike an average of their waking hours.

Up at 6 a. m.

"At a cost of about thirty dollars I installed a one hundred candlepower tungsten lamp and a two candlepower incandescent lamp in one chicken house and two sixty candlepower tungstens and a two candlepower lamp in the other. These I connected with switches in the house.

"As soon as the alarm clock goes off at six or a little after in the morning I turn on the switch and the chickens get up, thinking it is daylight. The lights are turned off at eight or eighty thirty, when it is full daylight and the neighbors' fowls are just arising.

"When it begins to get dusk, along about four, my daughter Dorothy, or my wife turns on the lights and they are kept going until nine at night, when I turn all out except the two candlepower lamps. These give just a sufficient amount of light to give the appearance of dusk, and the chickens begin going to roost. I leave the small lamps lit all night, so that if any of the chickens want to get up at night to eat they can do so.

Average Jumps from Twenty-six to Eighty-three.

'Eleven days after the lights were installed the daily average jumped from twenty-six eggs to eighty-three. During the moulting season under the old custom, when most of the food was going to feathers instead of eggs, I got only eleven eggs a day. Now I get fifty-two a day during the moulting season. It is merely an experiment in efficiency, and I hope to improve on it."

"Chickens think," said Mr. Newell. "If they know they are going to get plenty of food the next day they'll lay. By my method I keep them thinking they are getting the same amount of daylight all the year around, and I'm keeping them thinking all the time."

This article was either copied, or a new article made up from it, in the press in many papers throughout the United States. A clipping sent the author from Dunkirk, New York, had reproduced the pictures in the "Tribune" and as an insert picture had the following:

The following was given me by a friend as having been taken from the "San Francisco Call:"

"CRUELTY TO HENS IN DARKEST CHICAGO."

"They Have to Get Up at Six O'clock in the Cold Winter Mornings.

There is an ingenious gentleman in darkest Chicago who takes about the meanest advantage of his hens that has come to our attention. Under the solar conditions obtaining in Chicago, midwinter dawn is not due until about eight o'clock in the mornng, and so a natural hen is not supposed to quit her downy roost until that hour. But this Chicago man has equipped his hen house with electric lights. These he switches on at six o'clock in the morning.

The poor hens are aroused by the glare of light and their consorts loudly crow to hail the dawn of artificial day. Down from their roosts troop the fowls and straightway they make for their nests. With some twelve hours of light before them, the hens busy themselves in laying, and every hen does her duty once every three days, which is a better egg laying average than obtains when there is no electric light inducement to laying.

As natural darkness comes over Chicago, the gentleman turns on the electric light, and until nine o'clock the poor hens are kept awake, under the delusion that it is still daylight.

Such cruelty to the hens should be punished. The man ought to be forced to eat a dozen eggs every day."

The "Electrical World" of February 6th, 1915, had the following:

"EGG PRODUCTION INCREASED BY ELECTRIC LIGHT"

"According to the testimony of Mrs. George G. Newell, of *Brookfield, Ill., a suburb of Chicago, the effect of using artificial light in her chicken house to simulate the long days of summer has been the trippling of the egg output of her hens. In their tungsten-lighted compartments these estimable chickens now average one hundred fifty eggs per hen per year. A total of 18,000 eggs was produced in the Newell coops last year.

The increase in the productiveness of the hens has resulted, it is explained, from the duplication of summer lighting conditions during the dark days of winter. It was Mrs. Newell's theory that the hens did not lay many eggs during the winter months because they spent more time on their roosts and had less opportunity for scratching about for food. At a nominal cost the electric service of the Public Service Company of Northern Illinois was extended to the hen house. Each of the two sections is provided with a two candlepower lamp and a one hundred candlepower cluster.

At 6 a. m. on dark winter mornings when the family arises the lamps are switched on in the coops. At once the feathered occupants are roused to the day's activity of scratching for food. After the appearance of daylight outside, the lamps are turned off. With the return of dusk in the late afternoon they go on again, and they continue to burn until eight o'clock, when all are turned off except the two candlepower units. These lamps give a low illumination, simulating dusk, and the hens at once prepare to go on their roosts.

---

*Brookfield, Ill., has three depots and three postoffices, called Hollywood, Brookfield and Congress Park.

Fifteen minutes later, when all are in place, the small 'dusk' lamps are extinguished and darkness reigns on the chicken house until the next morning.

Two weeks after the present lighting system was installed the daily egg output had risen from twenty-six to eighty-three, according to the owner. Fifty eggs a day are now obtained during the moulting season, in comparison with eleven eggs a day secured under the former artificial lighting conditions."

This is a very fair article, except for the error in the last line, where it gives the impression that other artificial lights were previously used.

Judging by the letters received, the "Chicago Tribune" article aroused a general interest. This article, and all the others which the author has seen, with the exception of the one here reproduced from the "Electrical World," show that the subject was treated in either a humorous or sarcastic vein; and left the inference to the readers that it was cruel, tricky, or humorous to subject the hens to a longer business day.

The economical importance, of the subject, seems to have been left to the reader's own ability to draw his own deductions, and to read between the lines.

Corroborations of Long Business Day Benefits.

The first, of whom the author has knowledge, to follow in experimenting with a longer business day was Mr. J. C. Kline, of Congress Park. He seemed to take some interest in accounts of the experiment and its results, but did not seem to get enthusiastic, until he saw the plant in operation at the end of the summer of 1914. He equipped his poultry house with electric lights, and has received full benefits since in the supply of eggs in the fall and winter.

The next to follow was Mr. J. W. Allen, of Riverside, Illinois, to whom electricity was not available at the time. The author procured three Air-O-Lanterns, which produced a 300-cp light, with a consumption of one quart of gasoline, for from twelve to fifteen hours. Mr. Allen got one of these lanterns, and after putting the same into use began to receive returns in eggs within three or four days.

Mr. Maurice L. Newell, a brother of the author, got another of these Air-O-Lanterns to try on his poultry farm in

Michigan. He had been skeptical right along about the efficiency of light—attributing the author's good egg yield to other causes. His egg yield in November, 1914, however, had dwindled to twenty-two eggs in twenty-one days, and on November 26th the author took him an Air-O-Lantern, which was installed in the poultry house. The following extract from the letter of November 21st from Maurice L. Newell, will give some idea of the condition of the flock:

"Some of those early moulters, you know they started to moult in July, well they came just up to the laying point—red large combs, etc., and in good flesh—one or two started to lay, then they quit and are now in their second moult around neck and losing tail feathers. Also some of the yearling hens have done the same, and while the pullets are looking well, and some have large lay-over combs, have so far not received an egg from them."

A few days later the author got word from his brother, that he was shipping eggs to Chicago in case lots.

Mr. William Trefzger next installed electric lights in his poultry house. On February 1st the author wrote him as follows:

"Congress Park, Ill., February 1st, 1915.
Mr. Wm. Trefzger, 8541 South Sangamon Street, Chicago.

Dear Sir: Knowing that you have kept poultry for a number of years, and that you have recently installed electric lights in your poultry house, I shall appreciate the favor if you will write, giving me the results obtained from your experiment.

I am enclosing a stamped and addressed envelope for your reply, which will greatly oblige,

Yours very truly,
GEO. G. NEWELL.

And received the following reply:

8541 South Sangamon Street, Chicago, Ill., February 3rd, 1915.
Mr. Geo. G. Newell, Congress Park, Ill.

Dear Sir: I have received your letter dated February 1st, 1915. You will remember my conversation with you about the middle of last December, in which I asked you what caused my pullets to suddenly stop laying, and your reply that it was due to the short days.

At that time I could not see it your way. Within a few days I spoke to you twice again on the same subject. On the latter occasion you volunteered the theory:

First, that, as I had told you, my pullets were laying well, and would probably have continued to do so, falling off gradualy as the winter progressed.

Second, that the sudden change to cold weather which fell below zero at our place on December 14th, made a sudden call on the systems of the pullets for extra nourishment to maintain heat and vitality—thus forcing a curtailment in egg production.

Third, that the weather remaining cold for several days, the pullets had not been able, on account of the short days, to get back into laying form; and probably would not do so until Spring unless their business day was lengthened.

Soon after, my house was wired for electricity, and I ran a wire out to the poultry house, and started the light January 1st, 1915. The pullets, in the meantime, had gone into a heavy moult.

The first effect noticed was that the moulting stopped immediately. Next there was a great improvement in the looks of the birds. Their combs began to redden and we began to get eggs the third day.

We have in our flock six hens and twenty-six pullets. We did not keep any record of eggs, but they increased two or three a day until the fifteenth, when my wife suggested that, as we were getting so many eggs, we had better keep account of them. I enclose a calendar for January on which we kept this account. You will see on the fifteenth we got 16 eggs and then 18, 16, 18, 21, 22, 15, 24, 24, 19, 24, 19, 23, 19, 21, and on the thirty-first, 20 eggs, which is a total of 319 eggs in sixteen days.

We had tried liberal feeding without results. This record is very good, especially so considering the cold weather. The thermometer read below zero several mornings, and fourteen below zero January 28th.

I have been keeping poultry over twenty years, and never had such surprising results from anything. My wife was overjoyed, as she loves her poultry, and the sudden change from a condition, when we thought the whole flock would have to be sacrificed, to one of health and vigor, with the addition of good laying, was very gratifying to us. We think it is wonderful the way your theory has proved out in actual practice.

Yours very truly,

(Signed) WILLIAM TREFZGER.

CHAPTER VIII

Conclusions

Trying it on the Ducks

In the foregoing pages of this little book, the reader has been given an insight into the problem of egg production, and poultry keeping in general, from the author's viewpoint.

The whole subject has been put in such a manner, and the author has tried to state his premises clearly, so that the reader was not asked to take anything for granted.

The whole book has treated the subject with reference to hens only as producers. The production of duck eggs has taken great strides in the last few years, especially with the Indian Runner ducks, which have been termed "the Leghorns of the duck family."

In the keeping of ducks, however, the author does not feel competent to speak from actual experience. By the process of deduction, it is reasonable to predict that the same general effects of a longer business day will apply equally well in the production of eggs from ducks as from hens; and, on this deduction, this question should be worth serious consideration by those engaged in this branch of the poultry industry. It is well known that the duck family are naturally more nocturnal in their habits than hens.

Heredity and Performance

The flocks which we use for heavy egg production must be produced from birds which are known to be good layers. The birds must come from vigorous stock in order to grow and thrive quickly into producers.

Some seem to be prejudiced against breeding from hens which have laid well during the winter; but, if heredity counts for anything, how are we going to produce good winter layers unless we breed from those which have produced eggs freely in winter? If hens do not lay during the winter, can we reasonably expect winter layers as a result of transmitted heredity from such hens? If hens do not produce well in the winter months, they cannot make up the deficiency necessary to qualify, as heavy layers, by any summer laying records.

It is a safe conclusion that heredity alone will not give us good producers. Heredity must be supplemented by good, well regulated, and liberal feeding of balanced rations, together with a plentiful and constant supply of good drinking water. Only in this way can we turn to full advantage the hereditary qualities, which were transmitted to our birds.

We must give our flocks good quarters in which to work, and must keep these quarters in a wholesome and sanitary condition, as a requisite to that health and vigor, without which we cannot expect or obtain satisfactory results.

"Playing electric light jokes on the chicks" will not make them "lay for it," unless we give them plenty of opportunity to make use of this light. A longer business day will not feed the hens. Many will jump to the conclusion that artificial light by lengthening the day, will make their hens produce. Artificial light will lengthen the day, but will not and cannot be of any practical benefit to anyone who tries this innovation, unless such a person is also a liberal provider of food for his flocks.

Egg Producing as a Business Proposition

The author would place the value of artificial light to a flock of 2,000 layers at not less than $800 per year in increased production. This light, however, must be bright enough to flood the houses with light—not a mere glimmer.

Nothing could be further from the author's intention than to have the reader of this little work jump to the conclusion, from what has been written in the foregoing pages, that **anyone** can engage in the poultry business, and make a living or a competence, out of the business of producing eggs.

Like every other business, this business requires study, and attention to business affairs; and to enter this business without the necessary ability, or qualifications, cannot help but prove disastrous.

Many enter this business as a result of dreams that won't and can't come true. This business has some advantages, however, over other businesses—chief among which is the fact, true in all civilized countries, that the market has not been, and is not likely to be, fully supplied with good fresh eggs at all seasons. What a bonanza would such a condition create in any manufacturing line, or in a wholesale or retail business.

To illustrate how quickly a large flock of hens which are not producing can eat into a bank account, we have an example on Pages 59 and 60, where the feed in October, 1913, cost $26.13, and the production in eggs was only $7.91. A large flock on this same basis, would produce a large balance on the wrong side of the ledger in a short time. Even millionaires would tire of extended experience of this kind. If no profits are made, they at least expect to get their money back.

To the generality of poultry keepers, in a small way, this would be the time when they would reason that they must cut down expenses and withhold the feed.

To do so, however, would be suicidal; because, if that was done, the flock would not only remain in the non-producing class all winter, but would still cause a necessary expense for feed. By liberal feeding at this time, the flock was enabled to turn the scale in the following month, and do well all winter thereafter.

Capital and Equipment for Large Flocks

By careful watching, and close observation, large flocks can be housed and cared for more economically than small flocks. The labor can be cut down, by labor saving devices for carrying feed, litter, droppings, etc.; and the houses can be so constructed that they may be easily subdivided, at pleasure, by placing swinging doors in the divisions, in such a manner as not to be a hindrance or impediment to free ingress or egress.

With automatic feeders in use, a grain conveyor could be so equipped, and without great expense, as to fill all the feeders, in succession, by the use of power machinery.

Such arrangements, with an automatic supply of water, would enable the poultryman to care for large flocks with a minimum of expense.

Only the actual expenses for material necessary for production have been gone into in the preceding chapters. The items of general expense, labor, interest on investment, etc., have not been gone into or taken up. These various items can be figured out to suit the needs and circumstances of each producer.

As to the capital necessary to engage in this business, that also is a matter of individual judgment and opportunity. Some successful poultrymen have been saved from disaster by lack of capital, this lack compelling a safe and sure pace while gaining their experience.

Instances could be cited of poultrymen starting out with large capital, minus a practical knowledge of what was before them, who used up their capital in gaining such knowledge, and gave up the ship in disgust at the results, or from inability to borrow more capital at the time when they had learned how to use it.

On the other hand, instances could be cited of those who have had to do without capital because they had it not, or could not procure it, but who have started in, in a small way, and attained the necessary working knowledge as they went along, and who, as a result, have grown into successful and prosperous poultrymen.

These varying results are accounted for when we consider that a poultryman must **know** his business, and if he has to learn the business, after engaging in it, a very large percentage of losses with a small flock amount to but a few dollars; whereas the same percentage on a large investment is a very serious matter.

Every man must do his own thinking, and if the reader's thinking has been stimulated toward producing better eggs and many more of them, to his own pleasure and profit, it will be all the satisfaction desired by

THE AUTHOR.